EXCEL® MANUAL

to accompany

Elementary Statistics

Eighth Edition

EXCEL® MANUAL

Johanna Halsey
Dutchess Community College

Ellena Reda
Dutchess Community College

to accompany

Elementary Statistics

Eighth Edition

Mario F. Triola
Dutchess Community College

Boston San Francisco New York
London Toronto Sydney Tokyo Singapore Madrid
Mexico City Munich Paris Cape Town Hong Kong Montreal

Reproduced by Addison-Wesley Publishing Company Inc. from camera-ready copy supplied by the authors.

ISBN 0-201-70459-5

3 4 5 6 7 8 9 10 PHTH 03 02 01

CONTENTS

Preface

Purpose of the Manual

This manual contains step-by-step instructions to help you familiarize yourself with the spreadsheet program Excel. Our primary purpose in creating the instructions is to help you become proficient with the features of Excel that lend support to working with data in your statistics class.

We know that it is impossible to cover every possible option in using the program, so we have chosen the techniques that have worked well for us and our own students. There are usually at least two ways to produce the same results. Our hope is that providing you with a solid set of step by step instructions, you will become comfortable enough with the program to begin to experiment on your own, and share your discoveries with other students in your class.

Layout of the Manual

Other than Chapter 1, this manual follows your text section by section. Almost all of the exercises worked through in the tutorial instructions are taken directly from that particular section in your text. At the end of each section you will find several exercises written to give you an opportunity to practice the technology skills introduced within that section. Our hope is that you will immediately employ the features you learn in Excel to help you with the exercises and projects that are presented in that section of your book. You will find that utilizing the technology to work on many exercises in the book will afford you the essential practice necessary to become proficient with the program.

Using Technology Wisely

Any software package has its own learning curve. You should expect it to take a certain investment of time and energy and regular practice to become comfortable with using Excel. The more regularly you commit to using the ideas presented in this manual, the more proficient and adept you will become with using this program when and where appropriate.

While there are many, many places that Excel can, and should be integrated into the course material, you also can benefit from doing some of the work without using any technology. To truly understand some concepts, you need to perform at least some of the computations by hand. Once you have a core understanding of an idea, the technology affords you a way to find answers quickly and accurately.

Technology Notes

The instructions contained in this manual are written for a PC. Keystrokes may vary for those using a Mac. Mac users should note that Ctrl + click has the same functionality as a right click for PC users and a single click has the same functionality as a left click.

Early in the manual, you will be asked to load the Data Desk®/XL (DDXL) Add-In that accompanies your textbook. This Add-In supplements Excel, providing additional statistical tools not included in Excel. For example you will use DDXL to construct boxplots, confidence intervals and to perform hypothesis tests.

You can find the data sets from Appendix B of the main textbook on the CD-ROM that accompanies your text. You can also download these data sets from the Internet at http://www.awlonline.com/triola.

Final Notes

We feel that the benefits of using Excel in a statistics course are vast. Many companies look for employees who are proficient with using spreadsheets. By learning how you can use Excel to support your work in statistics, you will simultaneously be developing a skill that is highly valued in the business world. Our hope is that this manual provides you with a relatively painless entry into the world of spreadsheets! We hope you enjoy your learning journey.

Johanna Halsey

Ellena Reda

CHAPTER 1: GETTING STARTED WITH MICROSOFT EXCEL

SECTION 1-1: INTRODUCTION

One of the most valuable computer programs used in business today is the spreadsheet. Spreadsheets are used to organize and analyze data, to perform calculations and to show relationships in data through various types of charts and graphs. Many computers come preloaded with Microsoft Office ®. Excel is part of the Microsoft Office Suite. It is a popular spreadsheet program and can be used by all skill levels.

The purpose of this chapter is to provide the student with an introduction to spreadsheets and to prepare you to use Excel in the study of statistics. The best way to learn the basics is to dive in. As you explore Excel you will notice that there are often several ways to perform the same task. This manual will highlight only one or two of those ways. However that should not prevent you from trying other ways or using a method you already are familiar with.

SECTION 1-2: THE BASICS

You may open the Excel program one of two ways:

1) double click on the **Excel icon** found on the desktop screen.

OR

2) from the **Start** menu – highlight **Programs** – highlight **Microsoft Excel** -then click on the left mouse button (Mac users just click).

When you start Excel, a blank worksheet appears. This worksheet is the document that Excel uses for storing and manipulating data.

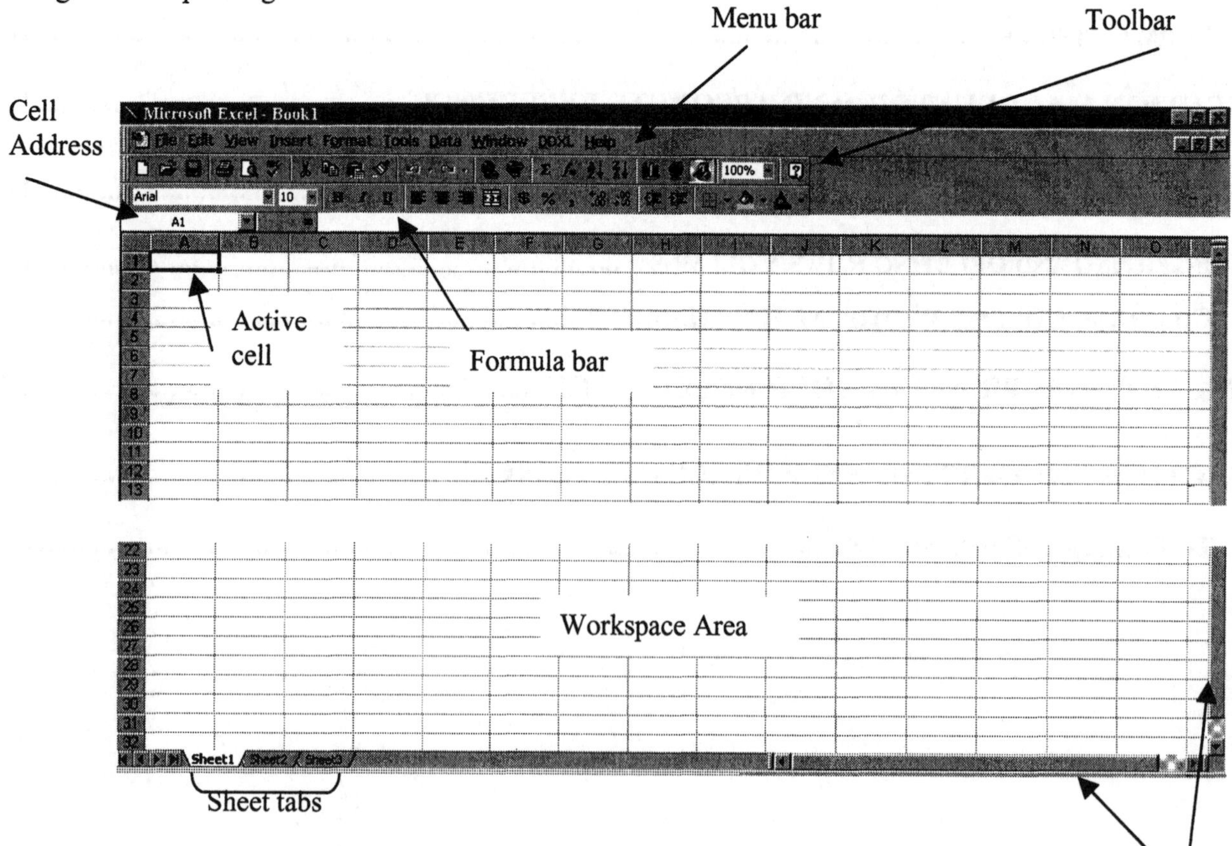

The following are some **basic terms** that you should be familiar with. Locate each of these on your Excel screen:

Toolbar – An area of the Excel screen containing a series of icon buttons used to access commands and other features. To find out what each toolbar icon does, place the mouse pointer over the button without clicking and its name will appear.

Menu bar – Groups of command choices. To view those choices click on one of the commands and a menu of commands in that group will drop down.

Worksheet area – The grid of rows and columns into which you enter text, numbers and formulas.

Cell – located at the intersection of a row and a column. Information is inserted into a cell by clicking on cell and entering the information directly.

Cell address – Location of a cell based on the intersection of a row and a column. In a cell address the column is always listed first and the row second so that A1 means column A row 1.

Active cell – The worksheet cell receiving the information you type. The active cell is surrounded by a thick border. The address of the active cell is displayed above the worksheet on the left.

Formula bar – Area near the top of the Excel screen where you enter and edit data.

Scroll bars – allow you to display parts of the worksheet that are currently off screen such as row 35 or column R.

Sheet tabs – Identify the names of individual worksheets.

SECTION 1-3: ENTERING AND EDITING DATA INTO EXCEL

When an Excel worksheet is first opened, the cell A1 is automatically the active cell. Notice that A1 is surrounded by a dark black box. This indicates that it is the **active cell.** Note that A1 is also shown in the **cell address** box as well.

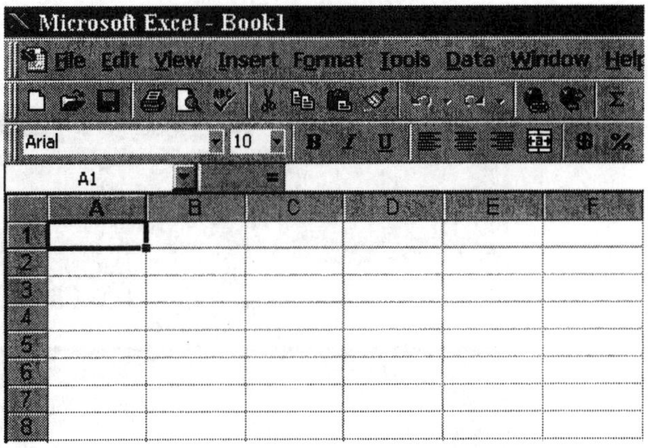

If you press a key by mistake, clear out information you may have wanted or find the screen isn't doing quite what you expect it to do try clicking the **Undo button** located on the toolbar. It looks like an arrow looping to the left.

THE FORMULA BAR

The formula bar is located in the fourth row of your Excel worksheet, to the right of the cell address box. The first window on the formula bar shown below indicates the cell address of the active cell. The **red X** is used when we wish to delete information we have typed into the active cell. The **green check mark** can be clicked to indicate that the data or formula you have entered into the active cell is acceptable.

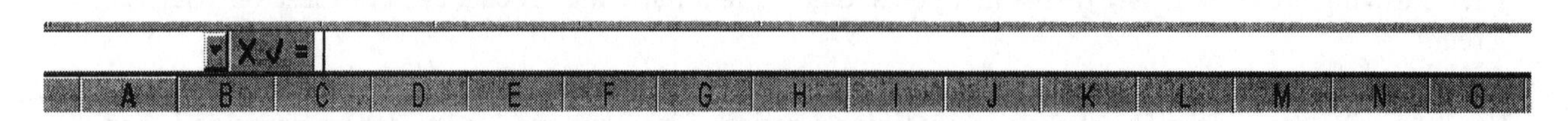

There are three types of information that may be entered into a cell:

- Text
- Numbers
- Formulas

ENTERING TEXT AND NUMBERS

The following exercise will give you some practice in entering both text and numbers into a worksheet.

1) Position the mouse pointer on cell C1 and click on the left mouse button. This makes cell C1 the active cell.

2) Type "ANNUAL SALES REPORT" (without the quotation marks) in cell C1 and press **Enter.** Notice that the active cell is now C2.

3) Activate cell A3 by clicking on that cell.

4) Type "REGION" in cell A3. Press **Enter**.

5) Similarly enter

   "QRTR 1" in cell B3
   "QRTR 2" in cell C3
   "QRTR 3" in cell D3
   "QRTR 4" in cell E3
   "TOTAL" in cell F3
   "EAST" in cell A5
   "SOUTH" in cell A6
   "WEST" in cell A7

6) To complete the worksheet shown below, begin by activating cell B5 and typing the data shown.

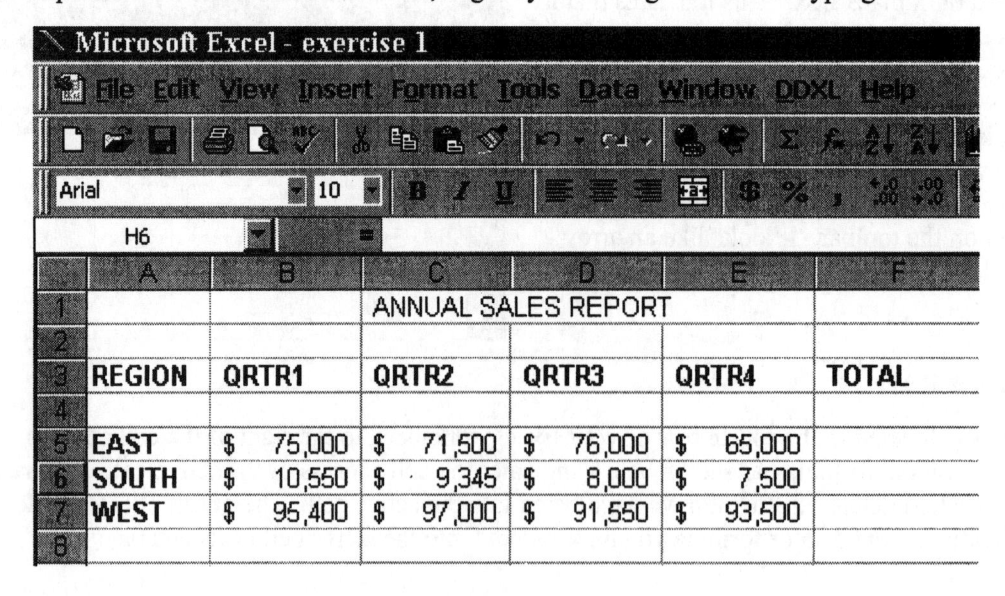

Note:
To move to the cell just below a cell you have entered data in press **Enter.**
To move to the right of a cell you have just entered data in press **Tab.**

EDITING INFORMATION

To empty the contents of a cell, we have already mentioned using the **red X** found on the formula bar. Information that has not been entered can be removed by using the backspace or delete key.

After data has been entered you can edit the information in a cell by activating that cell (click on it). The information in that cell is now displayed in the formula bar. Move your cursor to the entry you wish to modify or change. The cursor will turn into what looks like the capital letter I, commonly referred to as an I-beam. Place the I-beam at the point you wish to make changes, left click and proceed from there.

At this point you should save your work to a floppy disk or to the hard drive. It is a good idea to follow the adage "save and save often". We will come back to this problem after we have looked at the use of formulas in Excel to fill in the "TOTAL" column.

To save your worksheet on a floppy disk or hard drive

Do this by either clicking the **Save** icon on the toolbar (it looks like a floppy disk) or by clicking on **File** in the menu bar – and choosing the **Save** command.

- A dialog box will appear similar to the one you see on the right.
- Choose **Drive A** to save the information to a **floppy disk**.
- Choose **C** to save information to your **hard drive.**
- Enter a **file name** that will accurately indicate what the file contains and then click on **Save**.

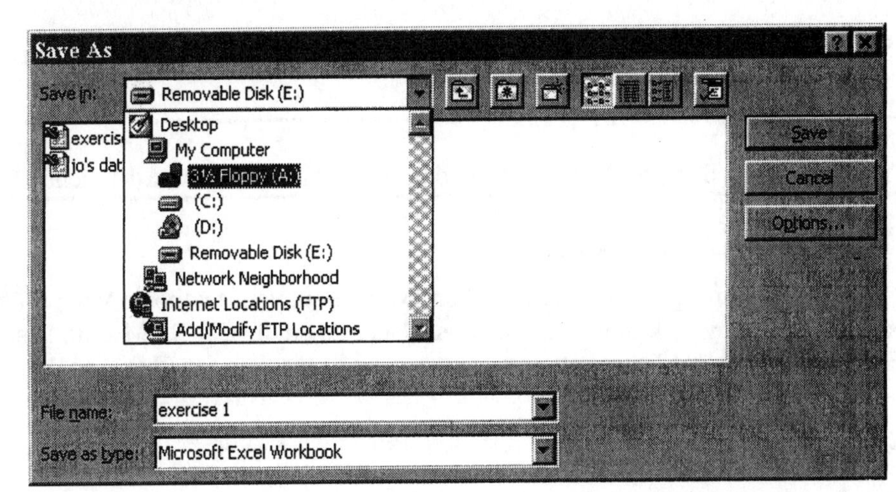

SELECT A RANGE OF CELLS

We often need to select more than one cell at a time. A group of selected cells is called a **Range**. To select more than one cell at a time

1) Click on cell B3, hold the left mouse button down.
2) With the cursor in the middle of the cell move down to cell B8.
3) Release the left mouse button. The cells B3 through B8 should now be highlighted.

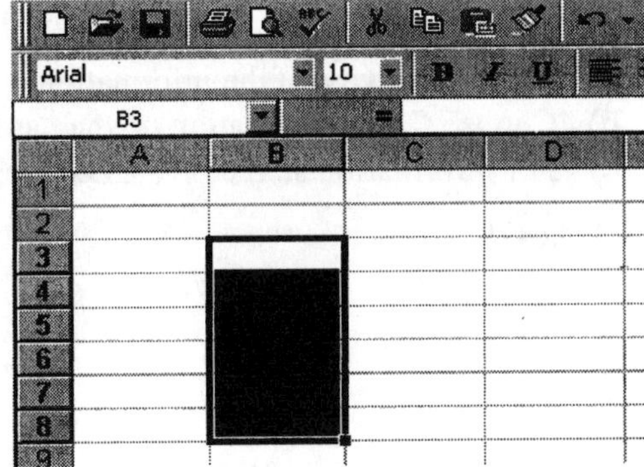

We refer to this **range of cells** as **B3:B8**, using a colon to separate the first cell in the range from the last cell.

DROPPING AND DRAGGING

To easily move or copy a range of cells from your worksheet we can use a shortcut method known as **drop and drag**.

To move a range of cells

1) Select the cells using the method previously outlined.

2) Hold the left mouse key down and drag the range of cells to their new location.

3) Release the mouse button.

To copy a range of cells

1) Select the cells as you did above, release the mouse.

2) Move the cursor to any edge until it resembles an arrow.

3) Hold the left mouse key and the **CTRL** key down at the same time.

4) Move to data to the location into which you wish it to be copied.

5) Release the mouse key , then the **CTRL** key. (the order in which you release the keys is important).

Note:
You can **Undo Drop and Drag** by locating this command in the **Edit** menu.

OPENING FILES IN EXCEL

As you work through this manual, and on exercises in your textbook, you will be asked to open data files saved to a floppy disk or hard drive or use data sets on the CD-ROM Data Disk that accompanied your Statistics textbook. These data sets are the same as the data found in Appendix B of your text.

To use the data sets on the CD-ROM

1) Open Excel.
2) Place the CD into the **CD-ROM drive** on your computer. This is often drive D or E.
3) Click on **File**, found on the menu bar, highlight **Open** and click.
4) A dialog box similar to the one shown on the next page will appear. You will need to choose the appropriate location for the files you wish to open.
   a) Choose **Drive A** if the information is on a **floppy disk**.
   b) Choose **C** for information on the hard drive.
   c) For information stored on a CD choose **D** if this is the **CD-ROM drive**.
5) Click **OK.**

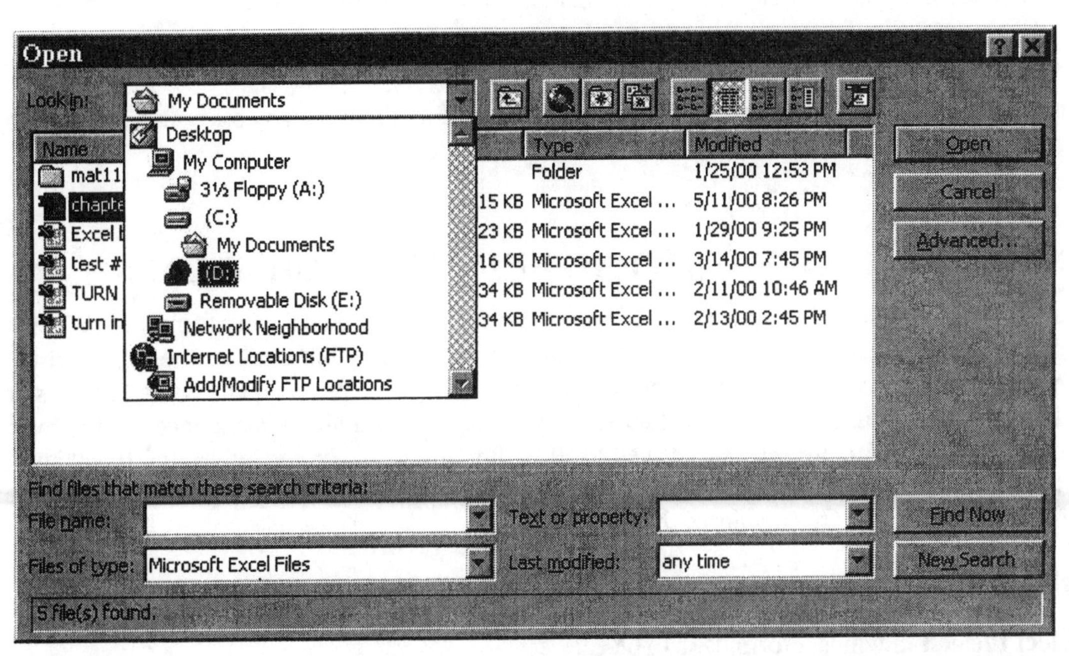

6) A second dialog box opens. This dialog box shows two files "DataSets" and "Software". Double click on **DataSets**.

7) This presents another dialog box as shown below. For a list of the files for Excel double click on the **Excel folder.** Use the scroll bar on the right of the dialog box to view the complete list of data files.

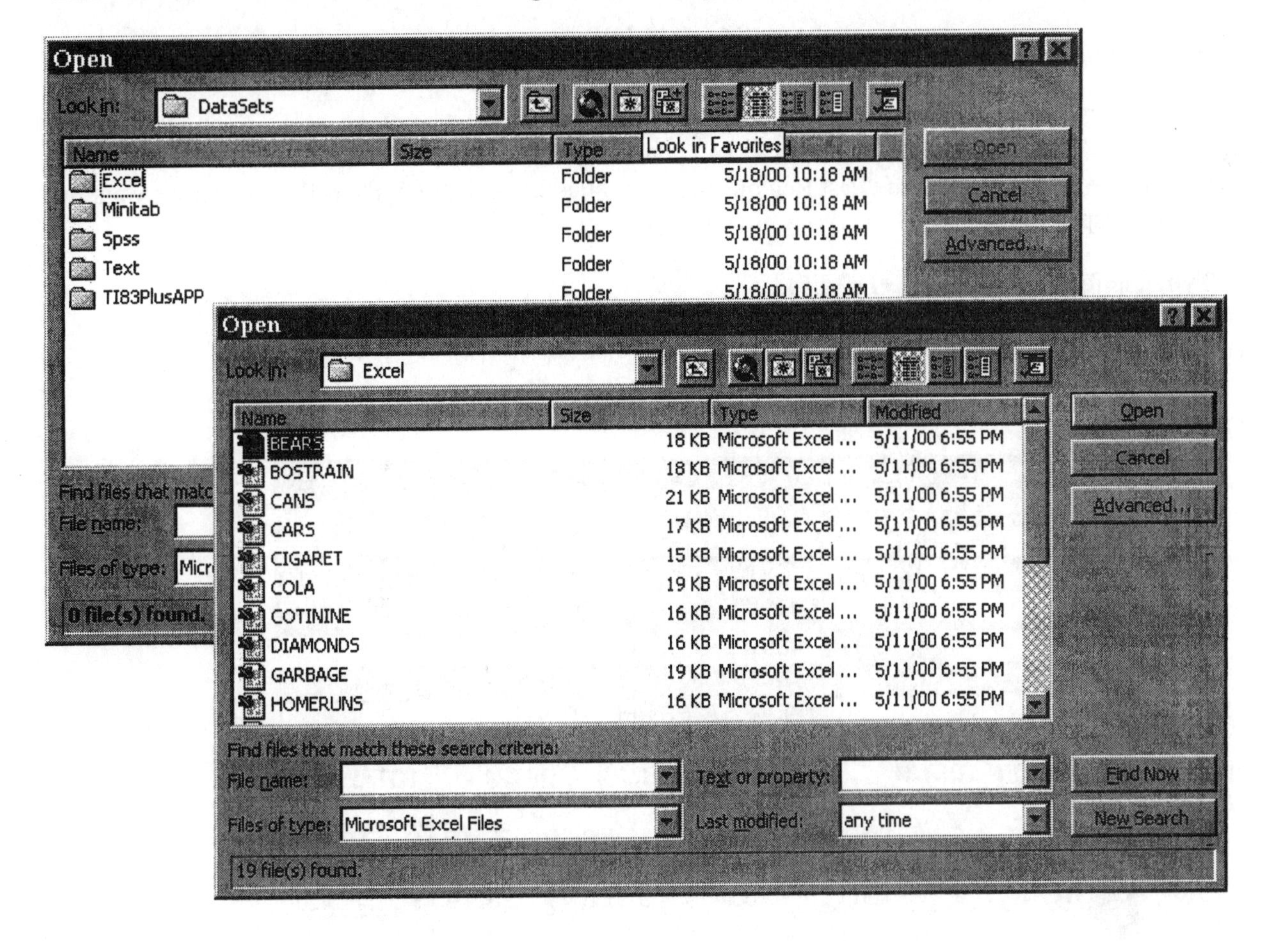

8) Identify the file you are looking for by file name. Click on the **file name**.

9) Click **Open**.

The same process is used when opening a file that contains data that you may save saved to a floppy disk or to your hard drive.

SECTION 1-4: UNDERSTANDING AND USING FORMULAS:

To analyze data in Excel it is frequently necessary to perform calculations using formulas. When entering a formula in Excel we can make use of either the actual value found within a cell or the cell address. When you type formulas in Excel the software program requires that you type an equal sign (=) at the start of your formula. This identifies the entry as a formula. As you create a formula, it is displayed in the formula bar. The numerical results are displayed in the cell itself. After you have entered a formula, Excel automatically performs the calculation. If you change a number stored in a cell address used in the formula, Excel automatically recalculates the results of that formula.

Entering a Formula

1) **Select the cell** in which a formula is to be entered.

2) Type an **= (equal sign)** followed by the formula. The formula should be surrounded by a set of parentheses. It is often easier to use the cell address within any formulas you write rather than the actual number although you can use either one. Using a cell address is advantageous when copying a formula to other cells.

3) After typing in the formula press **Enter**.

4) Open the file you saved when working in the previous section on text and numbers.

5) To find the total sales for the "East Region" begin by clicking on cell F5, the cell in which we want our answer displayed.

6) **Type** the formula =(B5+C5 +D5+E5).

7) Press **Enter**.

8) Repeat to find the totals of each row. Compare your answers with those below.

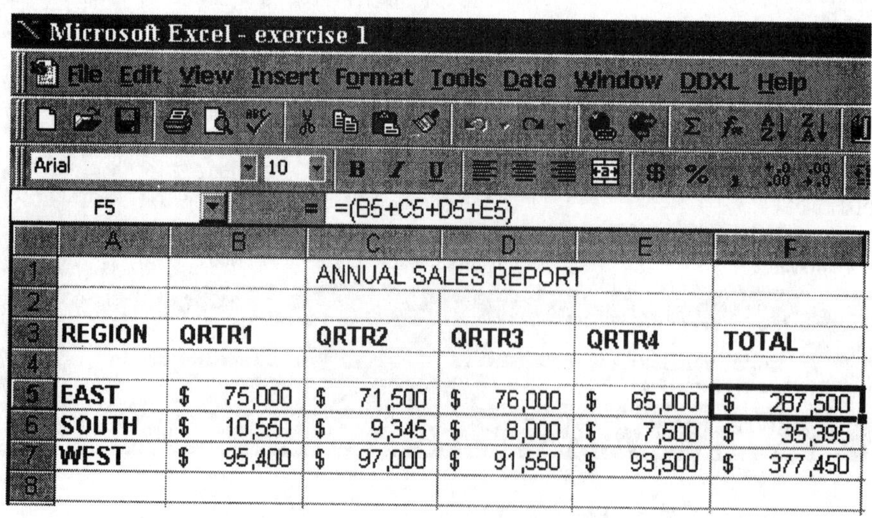

	A	B	C	D	E	F
1			ANNUAL SALES REPORT			
2						
3	REGION	QRTR1	QRTR2	QRTR3	QRTR4	TOTAL
4						
5	EAST	$ 75,000	$ 71,500	$ 76,000	$ 65,000	$ 287,500
6	SOUTH	$ 10,550	$ 9,345	$ 8,000	$ 7,500	$ 35,395
7	WEST	$ 95,400	$ 97,000	$ 91,550	$ 93,500	$ 377,450
8						

Notice that F5 has a black border around it indicating that it is the active cell. Also notice that the formula used to find the total in F5 is displayed in the formula bar.

Note:
Typing in each cell address to be added together in each of the columns above works well if there is a small number of cell addresses to be entered. If we had 50 data entries to add together in column B we would not want to type =(B1 + B2 + B3 + ….+ B50).

To save time we can enter the same information by typing = SUM(B1:B50).

Operators & Order of Operation

The formulas used in Excel make use of basic arithmetic operations. The symbols the are used for these **common operators** are:

Addition	+	Subtraction	—
Multiplication	*	Division	/
Exponents	^		

Excel also follows the basic **rules for order of operation** that we are all familiar with. Therefore it is important that when you input a formula you are very specific in the way the information is typed.

$\frac{2+3(2)^5}{5}$ would be entered as follows (2 + 3 * 2 ^ 5)/2 in Excel. Remember that formulas and equations generally contain a cell address rather than actual numbers.

Copying a Formula

In our previous problems, you probably retyped the formulas to find TOTAL in the appropriate cells. However, it is not necessary to re-type similar formulas. It is possible to copy an existing formula into other cells.

Method 1: (useful if you are copying the same formula to a series of cells)

1) Click on the cell that already contains the formula you want to copy.

2) Place your mouse on the lower right hand corner of the highlighted cell. When your cursor changes to a cross (commonly referred to as the **fill handle**), click and hold the left mouse button and drag the box to cover the cells where you wish to copy the formula.

3) When you release the mouse button the formula will be copied and adjusted for these cells. This is illustrated below.

F5 = =(B5+C5+D5+E5)

	A	B	C	D	E	F
1			ANNUAL SALES REPORT			
2						
3	REGION	QRTR1	QRTR2	QRTR3	QRTR4	TOTAL
4						
5	EAST	$ 75,000	$ 71,500	$ 76,000	$ 65,000	$ 287,500
6	SOUTH	$ 10,550	$ 9,345	$ 8,000	$ 7,500	
7	WEST	$ 95,400	$ 97,000	$ 91,550	$ 93,500	
8						
9						

Notice that the mouse pointer changes to a cross

Method 2: (useful if you are only copying one piece of information to another cell)

1) Click on the cell that already contains the formula you want to copy.

2) Click on the **copy** icon found on the tool bar (or use **Ctrl + C**).

3) Click on the new cell into which you wish to copy the formula.

4) Choose the **paste** icon found on the toolbar. (or use **Ctrl + V**).

Try both methods on the worksheet you have created to determine which works best for you.

SECTION 1-5: FORMATTING NUMBERS AND CELLS

The Excel worksheet that we have been working with in Sections 1-3 and 1-4 present information that represents the total sales figures. These are dollar amounts although this may not be apparent when you first enter the data and perform calculations.

The following shows the formatting icons found in your Excel worksheet.

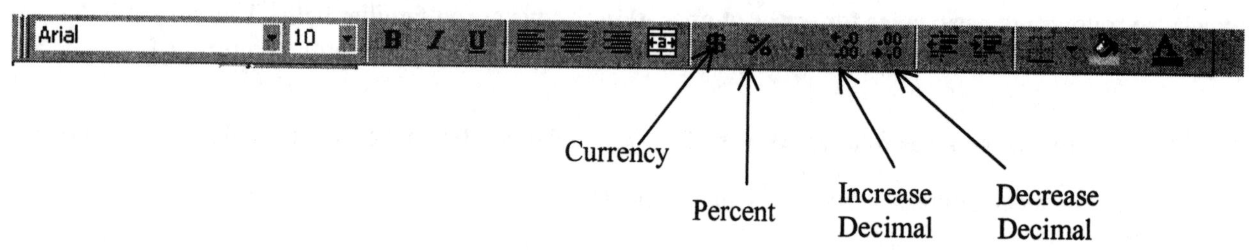

The **percent** and **comma** options are to the right of the **currency** icon. Try them and note the changes that are made.

The two icons to the right of the comma allow you to **increase or decrease** the number of places to the right of the **decimal** that are to be displayed. Try these on the data you have marked by highlighting the column as before and clicking on the desired icon.

Open the file you have been working on throughout this chapter.

FORMATTING CELLS

1) Click on the **B** at the top of the second column. The entire column is now highlighted.

2) Click on the **currency** icon ($) found on the formatting tool bar.

3) All data in that column should now contain a dollar sign in front of it.

Formatting a Range of Cells

It is possible to format a range of cells:

1) Select the cells you wish to format.

2) Click on **Format** from the Menu Bar, highlight **Cells** and click.

3) This will open a **Format Cells** dialog box. This dialog box presents a number of options to you, that include the ability to choose the type of number and the number of decimal places you wish displayed.

4) Try using this method to change all of your data in your worksheet to dollar amounts at the same time.

SECTION 1- 6: RELATIVE AND ABSOLUTE REFERENCE

Cell references can be **relative** or **absolute.** The cell references we have used so far are all **relative references** and when these references are copied to a new location they change to reflect their new position. In the preceding problem the formula from cell F5 was =(B5 + C5 + D5 + E5). When this formula was copied to cell F6 the formula changed to =(B6 + C6 + D6 +E6) reflecting a new relative position one row below where the original information was entered. This relative address feature makes it easy for us to copy a formula by entering it once in a cell and then copying its contents to other cells.

An **absolute reference** does not automatically adjust when moved to another cell and is used when it is necessary to retain the value in a specific cell address when copying a formula. In an absolute reference both values in the cell address are preceded by a $. For example, the formula =($B$5 + $C$5 + $D$5 + $E$5) will remain unchanged regardless of the cell to which it is copied. The dollar sign does not signify currency but rather is used to identify that the cell is an absolute reference.

Formulas can contain both relative and absolute cell references. For example, suppose I wished to determine what percentage of the total sales came from each region (East, South and West) in the problem we have been working with so far. I would begin by totaling the amount of sales found in column F. This information can now be found in F8 as seen below. To determine the percentage of sales for the East region I would need to divide the sales for the East region by the total sales OR F5/$F$8 and enter this formula in cell G5. The use of an absolute reference will allow us to copy this formula down column G. The top value will adjust to reference the cell we are in while the bottom value (the absolute reference) will remain constant, as we would want it to.

	A	B	C	D	E	F	G	H
1			ANNUAL SALES REPORT					
2								
3	REGION	QRTR1	QRTR2	QRTR3	QRTR4	TOTAL		
4								
5	EAST	$ 75,000	$ 71,500	$ 76,000	$ 65,000	$ 287,500	=(F5/$F$8)	
6	SOUTH	$ 10,550	$ 9,345	$ 8,000	$ 7,500	$ 35,395		
7	WEST	$ 95,400	$ 97,000	$ 91,550	$ 93,500	$ 377,450		
8						$700,345.00		
9								

SECTION 1-7: MODIFYING YOUR WORKBOOK

INSERTING AND DELETING ROWS/COLUMNS

Sometimes you will want to insert or delete columns or rows from your worksheet. If you want to create additional space in the middle of a worksheet you can insert a column or a row that will run the entire length or width of the worksheet. If you have an entire row or column that is no longer necessary you can delete the entire column or row.

To Insert a Row (or Column)

1) Highlight the row (or column) by right clicking on the number at the start of the row (or letter at the head of the column).

2) A dialog box should open. Choose **INSERT**. The new row will be inserted *above* the selected row. A new column will be inserted to the *left* of the selected column.

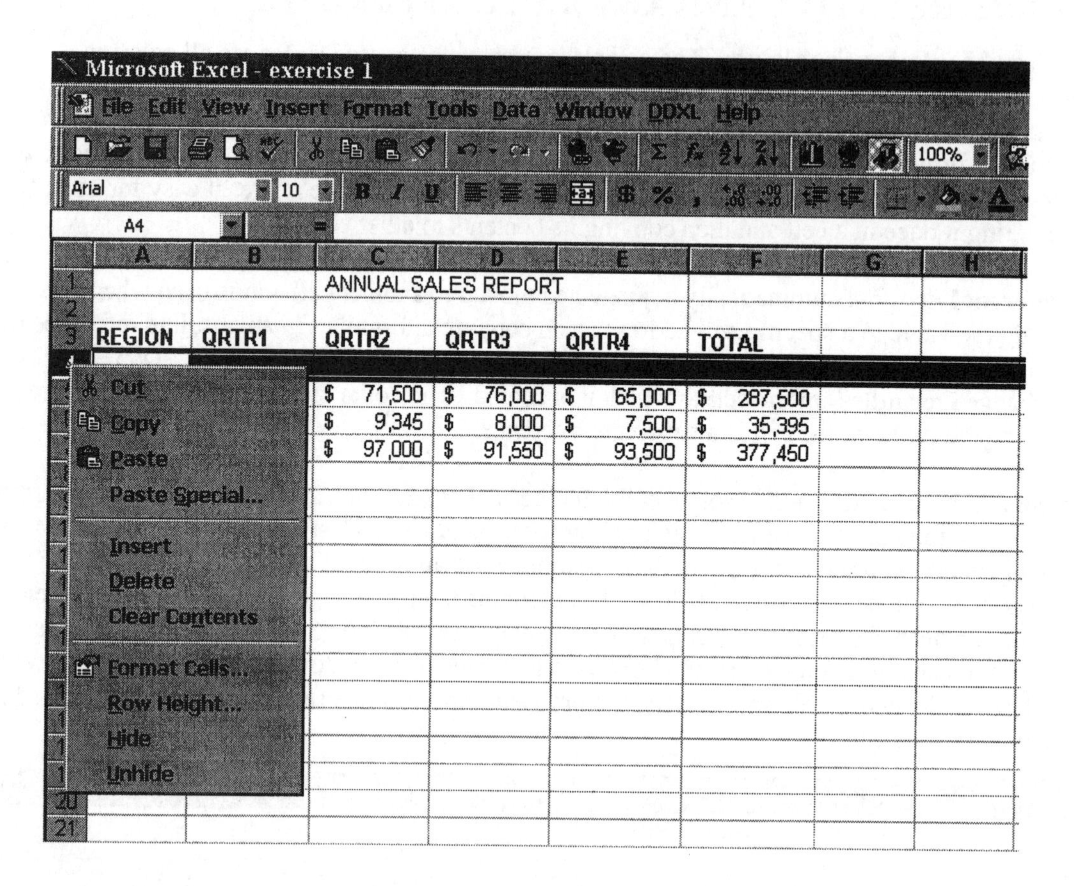

To Delete a Row (or Column)

1) Highlight the row (or column) by right clicking on the number at the start of the row (or letter at the head of the column).

2) As seen on the preceding page - a dialog box should open. Choose **DELETE** . The old row (or column) is removed from the worksheet and is replaced by the data in the adjoining row (or column).

Note:
Pressing the **Delete** key does not delete the selected row or column. It clears all data from the selection without moving in replacement data. Click a cell outside of the selected row (or column) to deselect it.

CHANGING COLUMN WIDTH AND ROW HEIGHT

Often we are in a position where we wish to change the width of a column. Text is often cut off because the column is not wide enough to display all that we have entered into the cell. If a cell can't display an entire number or date, the cell may fill with ####### or display a value in scientific notation. (Try entering a 12 digit number into a cell.)

To Adjust the Column Width

1) Position the mouse pointer on the right border of the lettered heading at the top of the column you wish to adjust. The mouse pointer should change to a black cross with an arrow head at each end of the horizontal line.

2) Press and hold down the left mouse button, dragging the right side of the column to increase or decrease the column width.

3) Move the mouse back and forth. Release the mouse button when you have a column the width you like.

To Adjust the Row Height

1) Position the mouse pointer on the lower border of the numbered row whose size you wish to adjust. The mouse pointer should change to a black cross with an arrowhead at each end of the horizontal line.

2) Press and hold down the left mouse button, dragging the border to increase or decrease the row height.

3) Move the mouse back and forth. Release the mouse button when you have a row at the height you like.

To Add Worksheets to a Workbook

When you open a new workbook in Excel you will see that there are three worksheets available to you. This gives you the ability to do three separate problems or variations of a problem all within one workbook. While only three worksheets are presented, it is possible to have up to sixteen worksheets within one workbook.

1) Click on **Insert** on the menu Bar,

2) Highlight **worksheet**, then click.

You will notice an additional worksheet tab in the bottom area of the screen.

To Rename a Worksheet

Initially the worksheets in Excel are labeled "Sheet 1" "Sheet 2" and "Sheet 3". To help keep track of information found in the various worksheets in your workbook, it is often useful to rename your worksheets.

1) Right click on the worksheet tab that you would like to rename

2) Highlight **Rename** and click. The sheet tab should now be highlighted.

3) Begin typing the new name for this worksheet.

4) Press **Enter** when you have finished.

SECTION 1- 8: PRINTING YOUR EXCEL WORK

It is often desirable to print a worksheet or workbook you have created in Excel. While it is possible to print all of the sheets in a workbook in one operation it is highly recommended that you print each worksheet separately. It is *always recommended* that you use the **Print Preview** before printing anything. This will allow you to catch and correct mistakes before you print a page.

1) To preview your worksheet before you print it, click on **File** from the Menu Bar, highlight **Print Preview** and click. This will give you an opportunity to make sure all of your work is presented within the printable page.

2) Click on **Close** once you have previewed your worksheet.

3) To print click on **File,** highlight **Print** and click. A Print dialog box will appear and should be similar to the one shown below. In it you will see the name of the printer being used (yours) as well as other options.

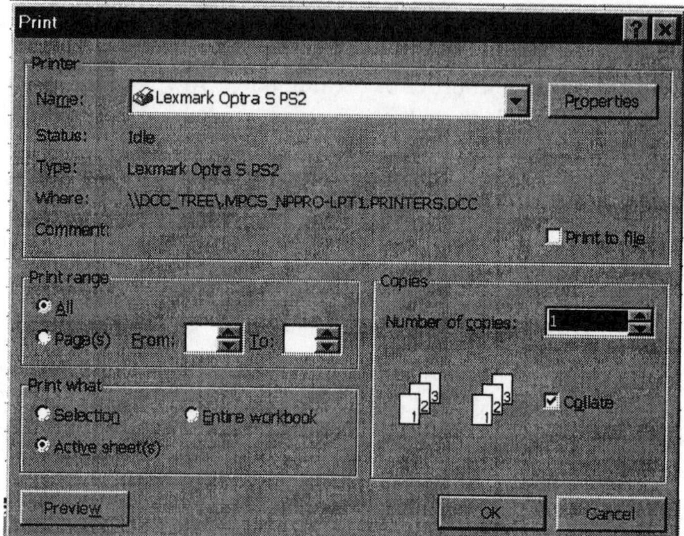

4) You can also customize your worksheet before printing. Click on **File**, highlight **Page Setup** and click.

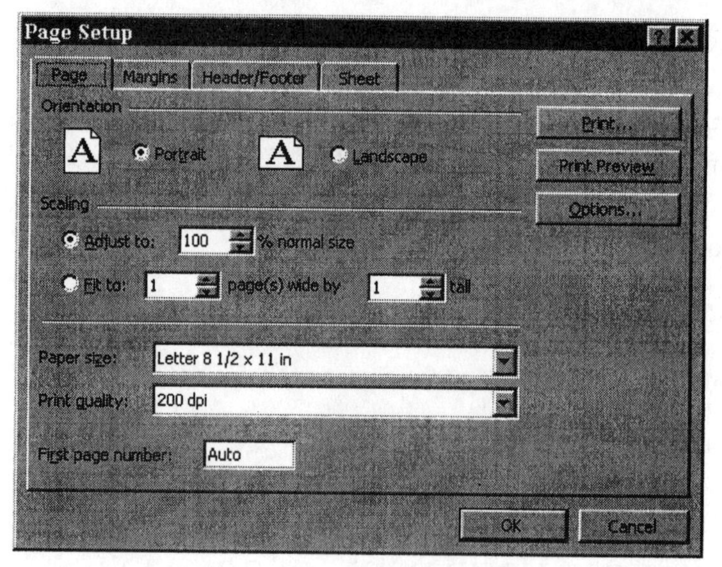

- **Page Tab** – allows you to choose the orientation of the printed page, either vertically (Portrait) or horizontally (Landscape). Landscape is useful for worksheets that have a greater width than they do length.
- **Margin Tab** – allows you to adjust the margin of the printed page.

- **Header/Footer Tab** – gives you the opportunity to include headers or footers on the printed page.
- **Sheet Tab** – the Print area is the most useful option on this tab.
- **Print area** - allows you to choose whether or not to include gridlines and row and column headings in the printout of your worksheet. It also gives you the opportunity to select a particular range to print from your worksheet. Click in the box to the right of Print area, highlight the range of cells you wish to print. You can also enter the information by typing in the range of cells.

Once you have completed making your choices click **OK**.

SECTION 1- 9: GETTING HELP WHILE USING EXCEL

Excel comes with a complete on-line Help feature designed to give assistance when you are having difficulty with a topic. You can access the Help system by selecting an option from the **HELP** Menu located at the top of the screen. Choose either Microsoft Excel Help or Contents and Index.

- **Microsoft Excel Help** gives you the opportunity to type in a question and search for an answer.
- **Contents and Index**:
  Contents is like the table of contents in a book providing an overview of major categories.
  Index is like the index in the back of a book providing an alphabetical list of the help that is available.

TO PRACTICE THESE SKILLS

It is important to practice those technology skills introduced in this chapter before moving on. To help you do this work through the following problem in Excel.

Temperature Conversions

The formula for converting degrees Fahrenheit to degrees Celsius is $C = \frac{5}{9}(F - 32)$. Use Excel to set up a spread sheet to do these conversions given a set of temperatures.

d) Type "Degrees Fahrenheit" in Cell A1.

e) Type "Degrees Celsius" in Cell B1.

f) In cells A2 through A11 enter the following temperatures recorded in degrees Fahrenheit: -10°, 0°, 10°, 32°, 45°, 50°, 68°, 75°, 83°, 95°.

g) In cell B2 enter the formula to convert from degrees Fahrenheit to degrees Celsius using an appropriate cell address.

h) Copy this formula through to cell B11.

i) Format the values in column B correct to 2 decimal places.

j) Rename your worksheet tab "Temperature Conversion".

k) Print out our worksheet.

CHAPTER 2: DESCRIBING, EXPLORING AND COMPARING DATA

SECTION 2-1: OVERVIEW

In this chapter, we will use the capabilities of Excel to help us look more carefully at sets of data. We can do this by re-organizing the data, creating pictures, or by creating summary statistics. The sections that follow take you through step by step directions on how to create frequency tables and histograms, as well as other types of visual models. We will also utilize the descriptive statistics capabilities of Excel to create representative values for the data set.

ADD-INS

You will be using two Add-Ins to Excel:

1) The **Data Analysis** Add-In, which can be added from your Excel software disk, and
2) **Data Desk/XL (DDXL)**, which comes with your book.

Instructions for adding these to your machine are listed below. Once you have added them to your machine, you will not have to go through these steps again.

Data Analysis

Analysis ToolPak is an Add-In in Excel, which may not have been loaded on your machine. If you do not see **Data Analysis** at the bottom of your **Tools** menu, you need to install this Add-In.

1) Click on **Tools**, and select **Add-Ins**.
2) In the **Add-Ins** dialog box, check **Analysis Tool-Pak**.
3) Click on **OK**.

The ToolPak will be loaded, and will appear at the bottom of the Tools menu.

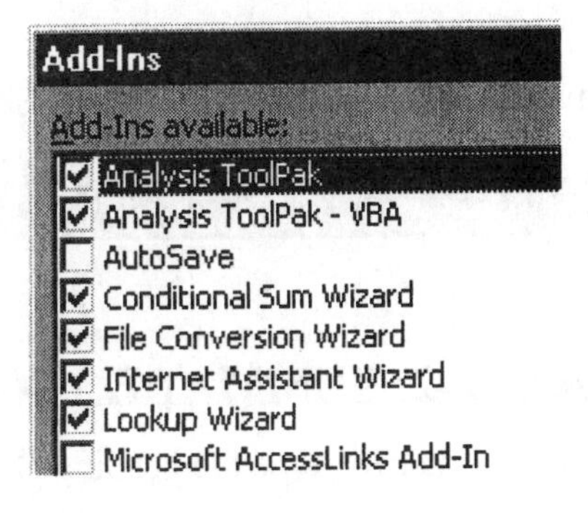

Data Desk/XL (DDXL) Add-In

Many of the applications throughout the book call for you to use the **DDXL** Add-In that is provided with your text book. To load this into Excel, follow the instructions below.

1) Begin by placing the DDXL CD into your computer. You should hear the CD running. The following screen should appear.

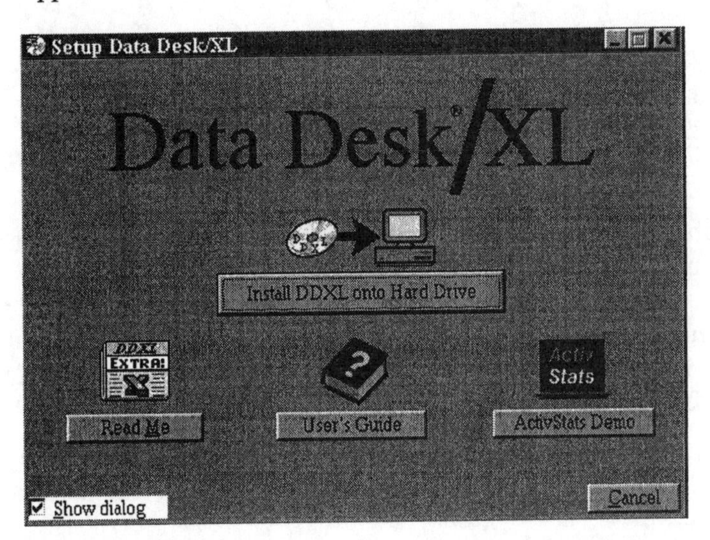

2) Click on **Install DDXL onto Hard Drive**. A dialog box will appear asking if you want to install DDXL onto your hard drive. Click on **Yes**.

3) Click on **Next** for the next 4 dialog boxes. Click on **Finish** in the final dialog box.

4) Recall that to add in some of Excel's extras we click on **Tools,** then click on **Add-Ins.**

If all goes well you will see **DDXL Add-In** in the **Add-Ins** dialog box. If not:

a) Click on **Browse**. Look in the **C** drive.
b) Double click on **Program Files.**
c) Double click on **DDXL.**
d) Double click on **DDXL Add-Ins.**

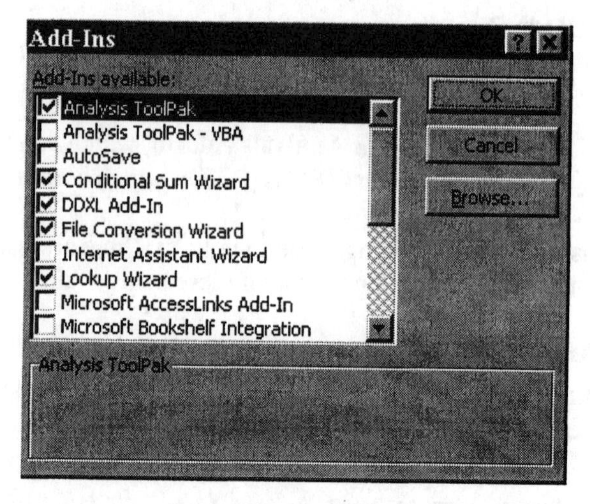

If you have successfully added **DDXL** to Excel you will see it featured on the toolbar as seen below.

SECTION 2-2: SUMMARIZING DATA WITH FREQUENCY TABLES

1) In a new worksheet, type the heading "Ratings" in cell A1.

2) Enter the data for Qwerty Keyboard Word Ratings into column A, starting with cell A2. (This data is in Table 2-1 on page 33 of your text.)

Qwerty Keyboard Word Ratings

2	2	5	1	2	6	3	3	4	2
4	0	5	7	7	5	6	6	8	10
7	2	2	10	5	8	2	5	4	2
6	2	6	1	7	2	7	2	3	8
1	5	2	5	2	14	2	2	6	3
1	7								

In order to create a frequency table, we need to indicate what upper class limits we want to use. Excel refers to these upper class limits as "bins".

If we want to construct a frequency table with 5 classes, class width of 3, starting with the value 0, we should enter the values 2, 5, 8, 11 and 14 into a column. We will name this column "Bin" to help familiarize ourselves with the terminology that is programmed into Excel.

3) In column C, type the name "Bins" in cell C1.
4) Type the values 2, 5, 8, 11, 14 into column C for the upper class limits, beginning in cell C2.
5) Click on **Tools** and select **Data Analysis**. (If necessary, go back to the instructions in Section 2-1 to learn how to add this feature to your machine.)
6) Double click on **Histogram** (or select **Histogram** and click on **OK**). You will see a dialog box similar to the following. To input the values shown, follow the directions below.

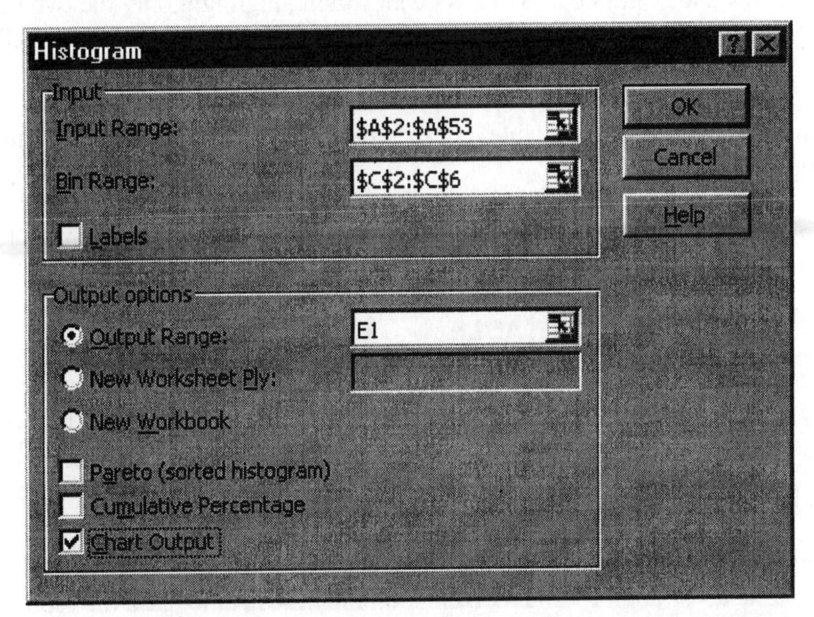

a) You need to tell Excel where the input is, and where the Bin values are. Since our input is listed in cells A2 to A53 we want to select cells A2 to A53 in the worksheet. In the **Input Range** box, type in "A2:A53", or select the data in the worksheet.

b) In the **Bin Range** box, type in "C2:C6", or select these cells in your worksheet.

c) To position your output in the same worksheet, click in front of **Output Range**, click inside the entry box, and type the cell location where you want your data to begin. Type in "E1", or select that cell in your worksheet. If you are going to overwrite data that already exists in your worksheet, a warning dialog box will appear.

d) To position your output in a separate worksheet, click in front of **New Worksheet Ply**, and type in a meaningful name such as Histogram.

e) Click on **OK**. You should now see a frequency table for your data.

Microsoft Excel - qwerty-chapter 2

	A	B	C
1	Bin	Frequency	
2	2	20	
3	5	14	
4	8	15	
5	11	2	
6	14	1	
7	More	0	

Note:
If the dialog box covers the data you want to locate, click on the symbol at the end of the line you want to enter information in. The dialog box will "collapse" into just one line, making it easier to navigate around your Excel worksheet. To reactivate the entire dialog box, you can either click on the symbol again, or press **Enter**.

You should remember to interpret the frequencies as the number of values that are less than or equal to the upper class limits that you typed into your bins, but greater than the previous upper class limit.

Looking at the frequency table for the Qwerty Keyboard Ratings, you should interpret that there are 14 ratings of either 3, 4 or 5 (less than or equal to the bin value 5, but greater than the previous bin value 2).

Modifying the Frequency Table

1) To remove the extra row and bottom bar in the frequency table, highlight only the two cells that you wish to remove. Click on **Edit**, highlight and click on **Delete**, click in the bubble by **Shift Cells Up**, then click on **OK.**

2) To rename the columns of the frequency table, click on the name at the top of the column, and type in the new name you would like to use.

3) Resize the column if necessary. You may want to refer back to instructions in Chapter 1.

CREATING A RELATIVE FREQUENCY TABLE

To create a relative frequency table, we need to add a column for relative frequencies to our frequency table. The relative frequency for a particular class will be equal to the number of values in that class divided by the total number of values in the data set. To set up this column in the frequency table follow the steps below.

1) In a cell under the entries in your bin column, type the word "Total".

2) In the corresponding cell under the frequency column type in the formula =SUM(F2:F7) and press **Enter**, or highlight cells F3 through F7 and then click on the **autosum** icon on the toolbar. You should now see the total number of data values in your data set, which should be 52.

3) At the top of the next column of your frequency table, type in the name "Rel. Freq".

4) In the cell immediately to the right of your data for the first class, enter the formula that will divide your frequency count for a class by the total number of data values. For example, if your first upper class limit is in cell E2, your frequency count for that class is in F2 and your total frequency count is in F8, you would type: =F2/$F$8, and press **Enter**. Notice that F2 is a relative address, while F8 is an absolute address, indicated by the $ signs.

5) Copy this formula to fill in the remaining cells in your table. You should see a table similar to the one below:

Qwerty Ratings	*Frequency*	*Rel Freq*
2	20	38.46%
5	14	26.92%
8	15	28.85%
11	2	3.85%
14	1	1.92%
Total	52	

Note:
You can represent your relative frequencies as either percents or decimals. To format your column,

a) Click on the column letter heading to select the entire column.
b) Click on **Format** in the menu bar, and click on **Cells**.
c) If you want decimal values, click on **Number**, and set the number of decimal places you want to be shown.
d) If you want percents, click on **Percentage**, and set the number of decimal places you want shown.
e) Click on **OK.**

TO PRACTICE THESE SKILLS

Apply the technology skills covered in this section by working through the following problems in your textbook. You can check the answers to the odd numbered questions by looking in the back of your text. Always make sure that you save your work, using a file name that will be indicative of the material contained in the Excel workbook. You will find you use many of the data sets more than once, and can utilize different worksheets within a file to separate the different work you do with the same data.

1) Load the data from Data Set 1 in Appendix B from the CD. You will find this data in the file named COLA.XLS. Use this data to complete exercises # 13 and 15 on page 41 of your text. You should do each exercise in a separate worksheet within a workbook, and name your worksheets with an appropriate name. This will help you remember where you can find the different parts when you retrieve your work.

2) Enter the data given in problem # 23 on page 42 of your text into an Excel worksheet. Make sure you label your column appropriately.

   a) For each set of data, construct a frequency table. You should use the same class width and upper class limits for both data sets. Use your frequency tables to answer the question posed in your text. Make sure you save your work with a name that is indicative of the material the file contains.

   b) For each set of data, construct a relative frequency table.

   c) Considering the frequency tables you constructed in part a, and the relative frequency tables you created in part b, which makes it easier to articulate your rational in answering the question posed in your text?

SECTION 2-3: PICTURES OF DATA

CREATING A HISTOGRAM

1) Using the instructions from section 2-2, follow the steps to create a frequency table. In addition make sure that the box beside **Chart Output** is checked.

2) Click on **OK.** You will see a screen like the following:

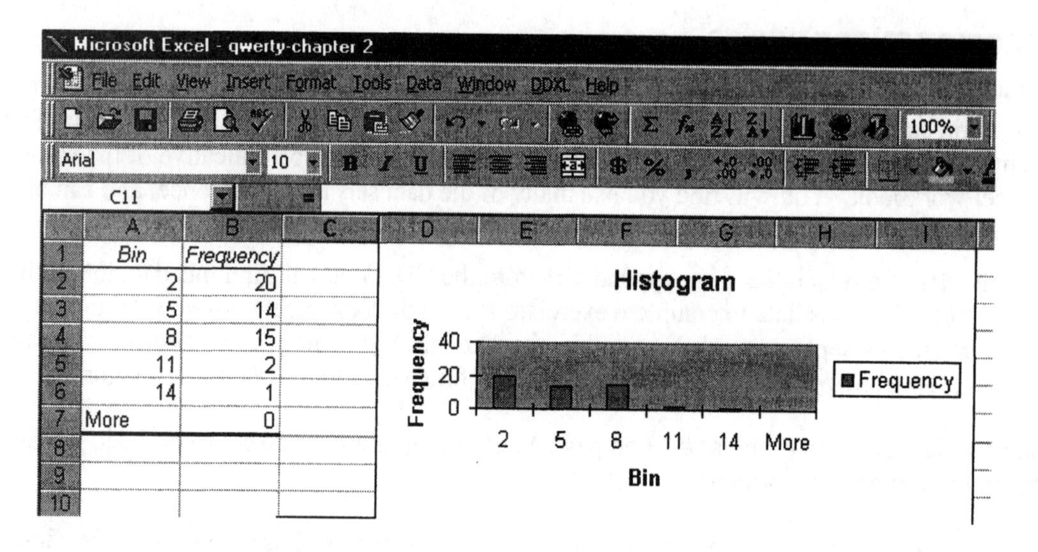

Note:
Excel uses bin numbers that are the upper class limits, but they are typically printed in the center of the histogram bars.

You will need to modify the histogram some to make it look the way you want it to.

3) You will want to remove the line that indicates "More". Highlight the two cells that you wish to remove.

4) Click on **Edit**, click on **Delete**, click in the bubble in front of **Shift Cells up**, then click on **OK**.

5) Follow the steps outlined below to further modify the histogram.

Removing the Gap Between Bars

1) To remove the gap between the bars on the Histogram, right click on one of the histogram bars.

2) Click on **Format Data Series** in the shortcut menu that is displayed. (You can also double click on one of the histogram bars to go directly to the Format Data Series dialog box.)

3) In the **Format Data Series** dialog box, click on the **Options** tab, and change the gap width value to 0. Click on **OK**.

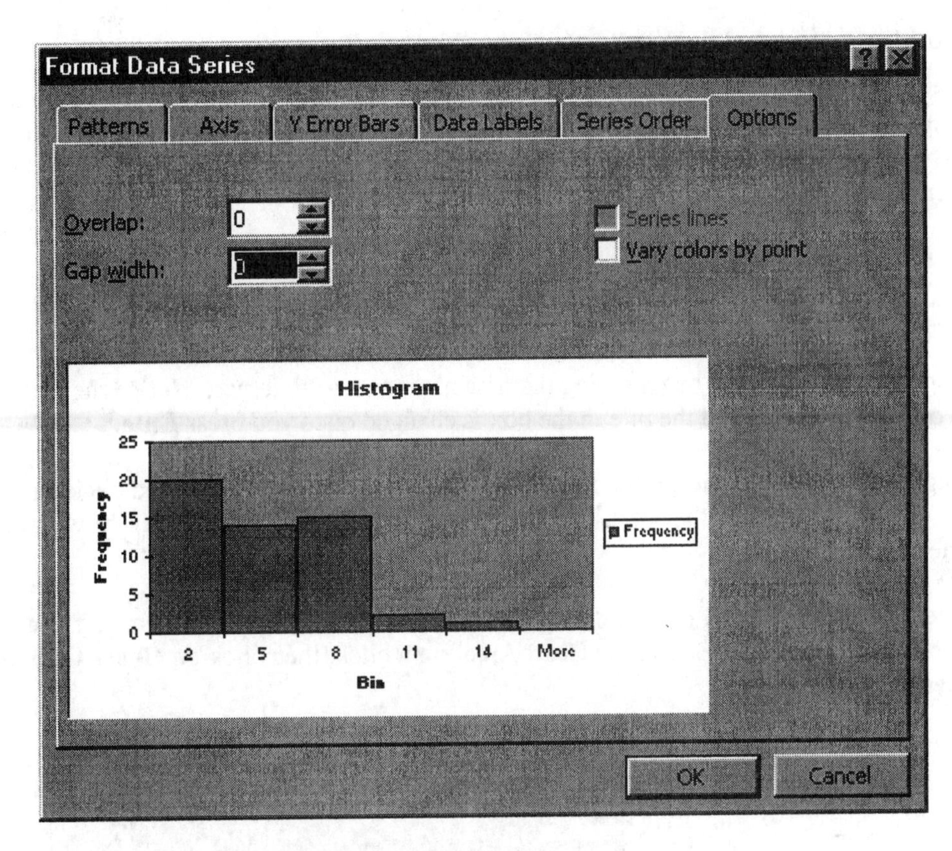

Selecting Regions & Accessing Formatting Options

1) Position your mouse over different parts of the white box containing the histogram. You will see "tags" come up telling you what the various regions are called. Regions include:

    Category Axis (Horizontal axis under the histogram where the "bin" values appear.)
    Chart Area
    Value Axis Title (Currently reads Frequency)
    Plot Area
    Chart Title (Currently reads Histogram)
    Category Axis Title (Currently reads Bins)

2) Click on a region and notice that "handles" appear around that region. This indicates that you have "selected" the region.

3) Right click while a region is selected to access the formatting menu. You have many options open to you in terms of what type of formatting changes you can make. We encourage you to "play" with the various options to create your own individualized picture.

Resizing a Region

1) Select the region you want to resize.

2) Notice the black "handles" on the selection box.
   a) By clicking and dragging on the corner handles, you can resize the region in both directions.
   b) By clicking and dragging on the side handles, you can stretch or shrink the region horizontally.
   c) By clicking on the middle handles at the top or bottom, you can stretch or shrink the region vertically.

Changing the Titles

1) Select the title you want to change by clicking on it.

2) To change the name, simply begin typing the new name you wish to use. Notice the text is typed up at the top of the worksheet, and the title in the box is changed once you press **Enter**.

3) To change other characteristics, right click once you have selected the title, and click on **Format Axis Title.** You now can change the color, font and alignment. You should experiment with the options available.

4) To delete a title, simply click on it to create the "handles", and then press **Delete.** If you delete a title, and later want to re-create it, select the Chart Area, right click, then click on **Chart Options** and click on the **Titles** tab.

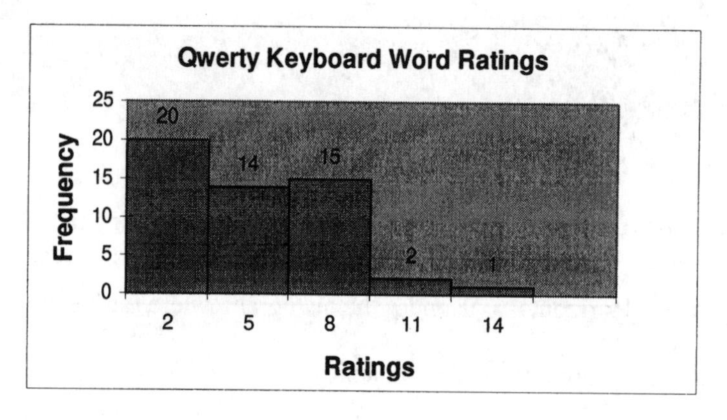

CREATING A RELATIVE FREQUENCY HISTOGRAM

Once we have the table showing relative frequencies, we can employ the Chart Wizard to generate a relative frequency histogram from this data. There are a series of 4 linked dialog boxes that you need to address when using the Chart Wizard.

1) Click on **Insert** in the menu bar and select and click on **Chart**, or click on the **Chart Wizard** icon on the tool bar.

2) In the **Standard Types** menu, make sure that **Column** is highlighted. Then click on **Next** at the bottom of the screen.

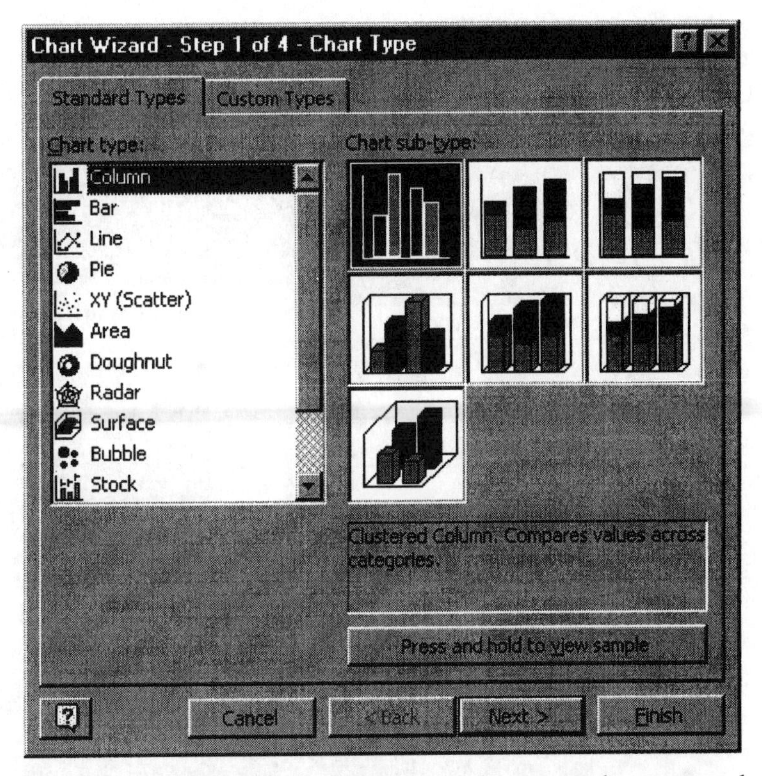

3) In the second dialog box (**Step 2 of 4 - Chart Source Data**) enter the range where your relative frequencies are listed by selecting these cells in the worksheet. Note that the word "Sheet1" (or the name of your sheet if you named it differently) comes up in the front of the range, and that the cell addresses are absolute addresses. **Unlike previous sheets, you cannot enter the range using just two cells with a colon in between.**

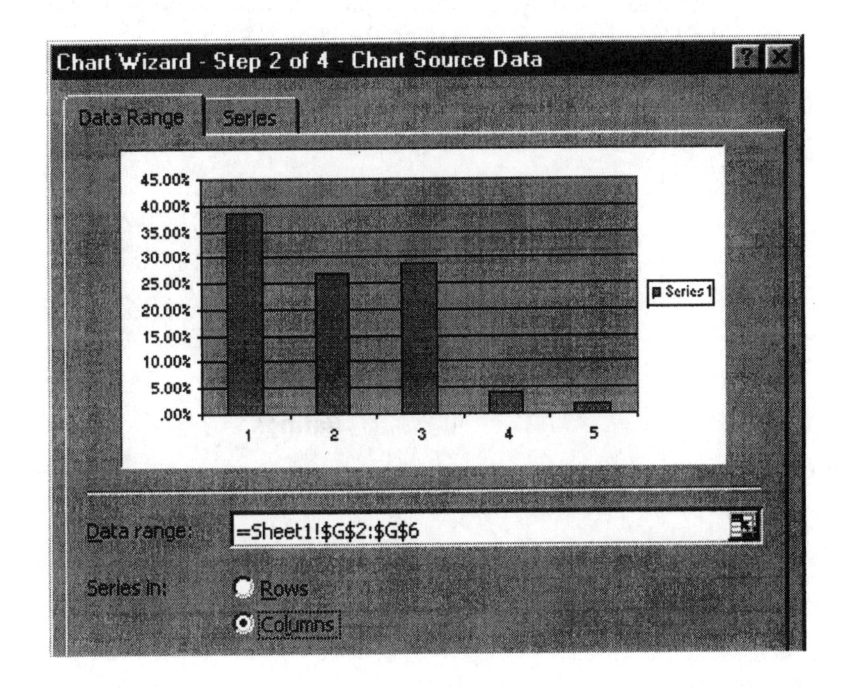

4) In this same dialog box, click on the **Series** tab, and in the **Category (X) axis labels** box, enter in the range where your upper class limits are listed by selecting the cells in your worksheet. Then click on **Next**.

5) In Step 3 of 4 - **Chart Options**, type in an appropriate Title, and descriptions of your x and y axes if desired. Click on each tab at the top of the dialog box, and determine which components you wish to select. Then click on **Next**.

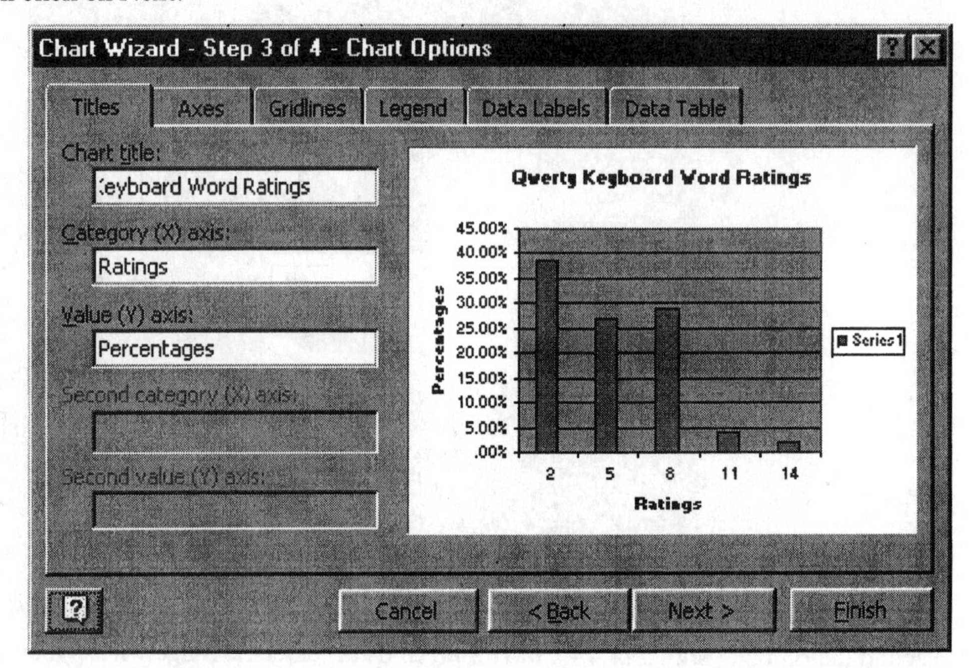

6) In Step 4 of 4 - **Chart Location**, determine whether you want your chart in a new worksheet (recommended option) or inserted in an existing worksheet. Then click on **Finish**.

7) You should see the basic relative frequency histogram. You will again want to modify the histogram by changing the gap width to 0, and by resizing the picture. See the previous instructions on pages 25 and 36 on Removing the Gap between Bars, and on Resizing a Region.

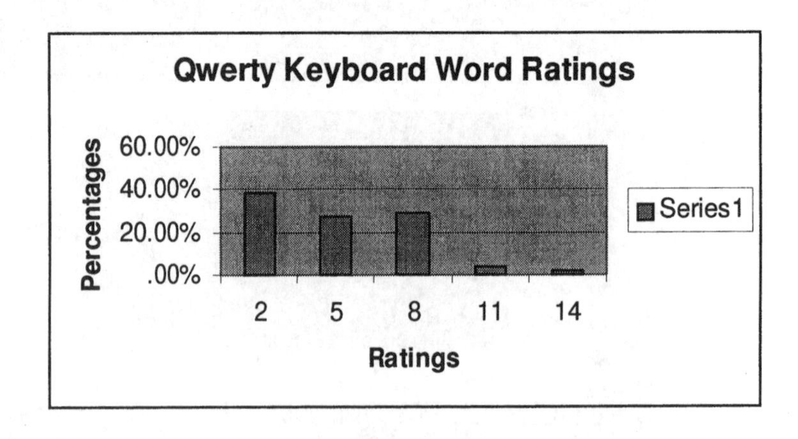

CREATING A FREQUENCY POLYGON

Once we have the frequency table for our data, we may choose to represent it graphically as a frequency polygon rather than a histogram. Since frequency polygons display points above the class midpoints, we first need to enter a new column for these values.

1) Copy your "Bin" and "Frequency" information to a new worksheet

2) Replace the values in your "Bin" with the class midpoints of 1, 4, 7, 10, 13.

3) Click on **Insert** from the menu bar, and click on **Chart.**

4) Click on **Line** under **Chart Types**, click on the first option in the second row of possible graph types, and press **Enter**.

5) In the **Data Range** box, enter the range where the frequencies are located by selecting them in your worksheet.

6) Click on the **Series** tab at the top of the dialog box, and in the **Category (X) axis labels** box, enter the range where the class midpoints are located by selecting them in your worksheet.

7) In Step 3 of 4 - **Chart Options**, type in an appropriate Title, and descriptions of your x and y axes if desired. Click on each tab at the top of the dialog box, and determine which components you wish to select. Then click on **Next**.

8) In Step 4 of 4 - **Chart Location**, determine whether you want your chart in a new worksheet (recommended option) or inserted in an existing worksheet. Then click on **Finish**.

CREATING A PARETO CHART

We can use Excel to sort qualitative data in order of decreasing frequencies, and then call on the **Chart Wizard** to create a Pareto chart.

1) Enter the following table into Excel:

Job Sources of Survey Respondents	Frequency
Help-Wanted	56
Executive Search	44
Networking	280
Mass Mailing	20

2) We first need to sort the data in order of decreasing frequency. Click on any one cell in the Frequency column. Do **NOT** select the entire column!

3) Click on the icon in the menu bar which shows descending from Z to A. You should see the information has been sorted in descending order, and the names associated with each value have stayed in the proper association.

Note:

If you had highlighted the entire column, only that column would have been sorted, and you would not have maintained the appropriate matching between the names and numbers.

4) Click on **Insert**, then click on **Chart**, and click on **Column.** Follow through the 4 linked steps in the Chart Wizard. **When you enter a range, you must do so by selecting the appropriate cells in your worksheet**. Notice that the word "Sheet1" (or the name of your sheet if you named it differently) comes up in the front of the range, and that the cell addresses are absolute addresses. Unlike previous sheets, you cannot enter the range using just two cells with a colon in between.

5) Enter your range where your frequencies are listed as the **Data range**.

6) Under the **Series** tab, enter the range where your categories are listed as your **Category (X) axis** labels.

7) Select appropriate titles and options under the **Chart Options** menu.

Modifying Your Pareto Chart

Once you have your basic picture, you can make further modifications to "professionalize" your chart.

1) Hold your cursor over any of the category names until you see the tag **Category Axis** appear. Double click once you see this tag,

2) You will see a **Format Axis** dialog box. Notice that one of the tabs is labeled **Alignment**. Click on this tab.

3) Position your cursor on the red diamond, and rotate the line all the way to the up position. Then click on **OK**.

4) You will see that your text has been aligned vertically.

5) Explore other options in terms of formatting the axes and the titles.

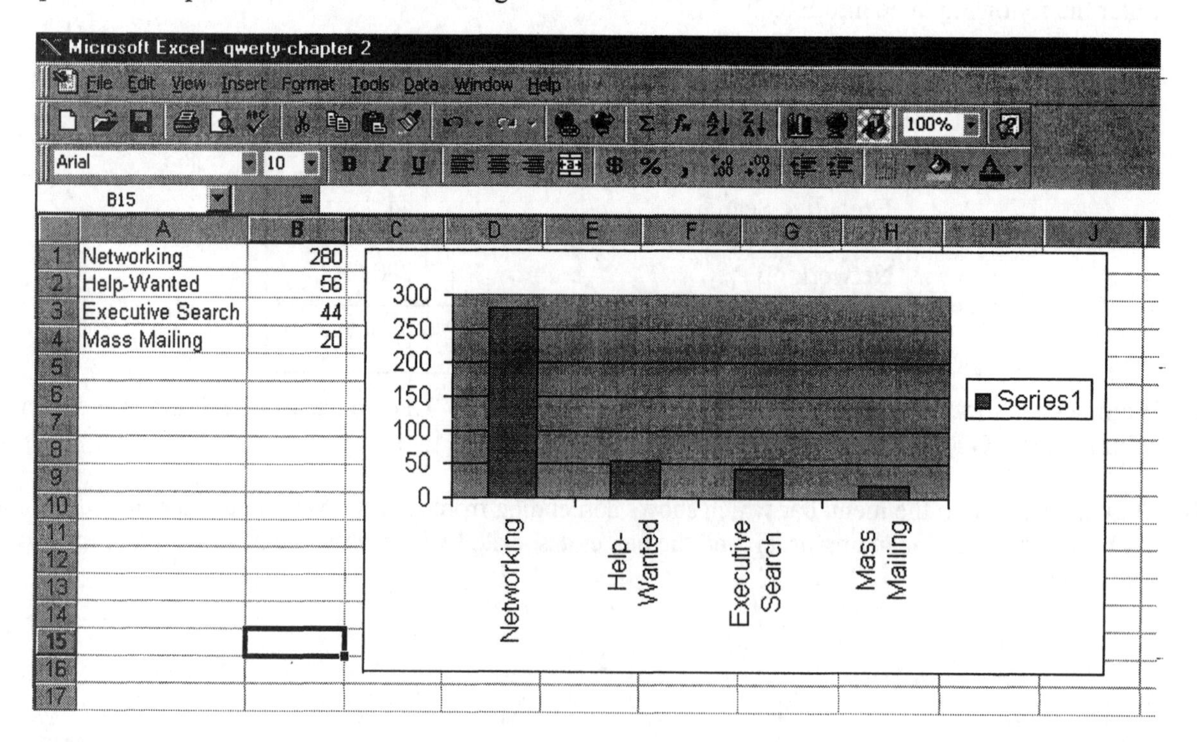

CREATING A PIE CHART

We may decide that we want to graphically show qualitative data on a pie chart, rather than as a Pareto chart. Let's work again with the information on Job Sources of Survey Respondents.

Job Sources of Survey Respondents	Frequency
Help-Wanted	56
Executive Search	44
Networking	280
Mass Mailing	20

1) Enter this information into two adjacent cells in a new worksheet, or copy it into a new worksheet if you have already entered it for the Pareto chart.

2) Highlight the cells with the sources and frequencies.

3) Click on **Insert,** then click on **Chart.** You will be taken through a series of 4 linked dialog boxes.

4) In the **Chart Type** box (Step 1 of 4) , click on **Pie**, and then select the type of pie chart you want in the **Chart sub-type** menu box. Then click on **Next** at the bottom of the screen.

5) The second dialog box is entitled **Chart Source Data**. Notice that your data range is already entered. Click on **Next**

6) In Step 3 of 4, you can enter your **Chart Options**.

7) Click on the **Legend** tab, and indicate where you want the legend box to appear in relation to your pie chart.

8) Click on the **Data Labels** tab, and select what type of labels you want to use for your pie chart. Try clicking in front of each to see how they will appear on the graph. Click on **Next**.

9) In Step 4 of 4, you can enter where you want your **Chart Location**.
   - Placing your chart in a new sheet is recommended. If you choose this option, enter in a descriptive name for the sheet, such as "Pie Chart".
   - If you want your chart in an existing sheet, enter the sheet name or number. Pressing the down arrow at the end of the entry box will show a drop down menu of the current sheets available in your workbook.

10) Click on **Finish**.

Modifying Your Initial Pie Chart

You will now see your initial pie chart. You should experiment with resizing and reformatting the various parts of your chart. It is recommended that you resize the pie itself so that it takes up more of the Chart Area.

1) Move your cursor close to the circle until you see the tag **Plot Area** appear. Click once you see this tag appear, and you will see a rectangle with handles surround your circle. You can now resize the circle within the **Chart Area**.

2) Move your cursor over the labels until you see the tag **Data Labels**. Click once you see this tag appear. You will see handles appear by the various labels. Right click, and in the menu box which appears, highlight and click on **Format Data Labels**. Experiment with the different options available.

3) Move your cursor over the legend until you see the tag **Legend**. Click once you see this tag appear. Right click, and then click on **Format Legend**. Experiment with the different options.

PRINTING A GRAPH

You can elect to print only the final graph that you create.

1) Select the graph you wish to print by positioning your cursor somewhere in the **Chart Area**, and clicking once.

2) Click on **File**, and in the drop down menu which appears, click on **Print Preview**.

3) You should now see a screen shot of what your graph will look like when you print it.

4) You can change the page margins by positioning your cursor on the dotted page guide you want to move. Your cursor should change into a double-headed arrow bisected by a straight line segment. Hold your left click button down and drag your cursor to the position you want the margin to be. Then release the mouse. You should see your page margins change in the direction you moved.

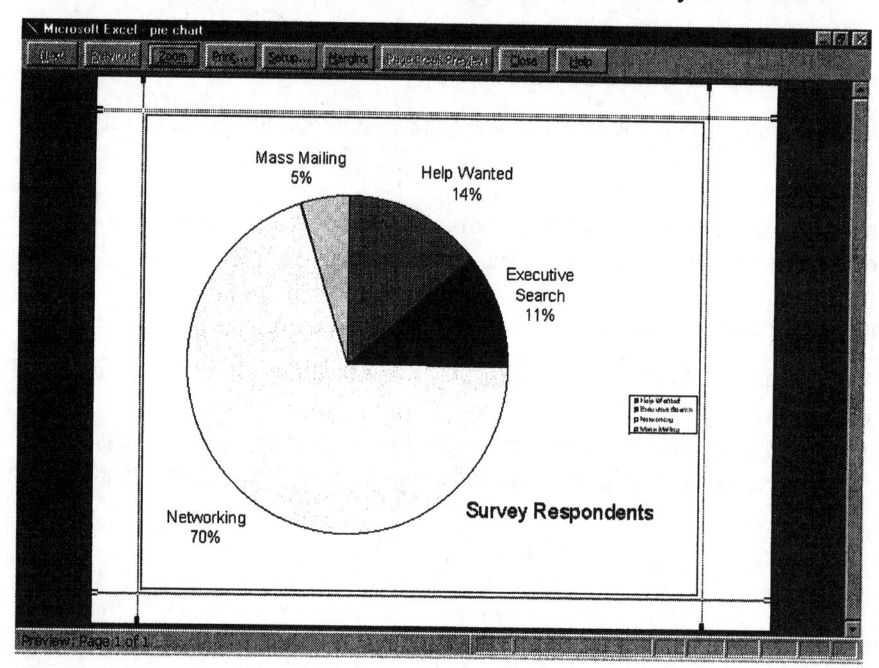

5) If you want to modify your graph, close the **Print Preview** window and make the changes you want to the graph.

6) Re-check your picture in the **Print Preview** window again. When your picture is as desired, press the **Print** key found at the top of the **Print Preview** window.

TO PRACTICE THESE SKILLS

You can apply the technology skills covered in this section by working through the following exercises. Make sure you save your work using a file name that is indicative of the material contained in your worksheets.

1) This exercise reviews creating **frequency polygons**. Open the Excel file where you constructed the frequency tables for exercises 13 and 15 from section 2-2. In this same workbook, complete exercise 10 on page 52 of your textbook. Make sure you save your expanded file.

2) This exercise reviews creating **relative frequency histograms**. Open the Excel file where you constructed the relative frequency tables for exercises 13 and 15 from section 2-2. In this same workbook, create two relative frequency histograms. Make sure you save your expanded file.

3) To practice creating a **histogram**, load the data from your CD for Data Set 12 in Appendix B (CANS.XLS) into a new worksheet. Use this data to complete exercise 29 on page 54 of your book.

4) To practice creating a **pie chart and pareto chart** complete exercises 21 and 22 in your text.

SECTION 2-4 & 2-5: DESCRIPTIVE STATISTICS - MEASURES OF CENTER AND VARIATION

Copy the Qwerty Keyboard Word Ratings Data to a new worksheet if you have previously worked with it, or enter the data into column A, starting with cell A2, and naming the column in cell A1. This data can be found in Table 2-1 on page 33 of your text.

Qwerty Keyboard Word Ratings

2	2	5	1	2	6	3	3	4	2
4	0	5	7	7	5	6	6	8	10
7	2	2	10	5	8	2	5	4	2
6	2	6	1	7	2	7	2	3	8
1	5	2	5	2	14	2	2	6	3
1	7								

Producing a Summary Table of Statistics

1) Click on **Tools** in the menu bar, and then click on **Data Analysis.**

Note:

If **Data Analysis** does not show up as an option, you need to load this as an Add-In in Excel. Follow the directions in section 2 - 1.

2) Click on **Descriptive Statistics.**

3) In the **Input Range** box, type in "A2:A53", or select these cells.

4) Click in the circle before **Output Range**, and then click inside the white entry box. Then type in "C2", or click in cell C2 in your worksheet.

5) Click in the box before **Summary Statistics**, and then click on **OK**.

6) You will see your **Summary Statistics** displayed starting in cell C2 of your worksheet.

7) Resize column C so that the full words can be seen.

8) Click on the title **Column 1** and type in "Qwerty Data Statistics".

9) You may want to delete some of the standard choices from your list. Select the two cells that contain the information on Kurtosis, and press **Delete**. Do the same for Standard Error.

10) You now have empty rows in your table. Select all the information in the two columns of your table. Do not include the title in your selection box . Then click on the **Sort** icon from A to Z. You will now see the information presented in alphabetical order, with the extra rows eliminated.

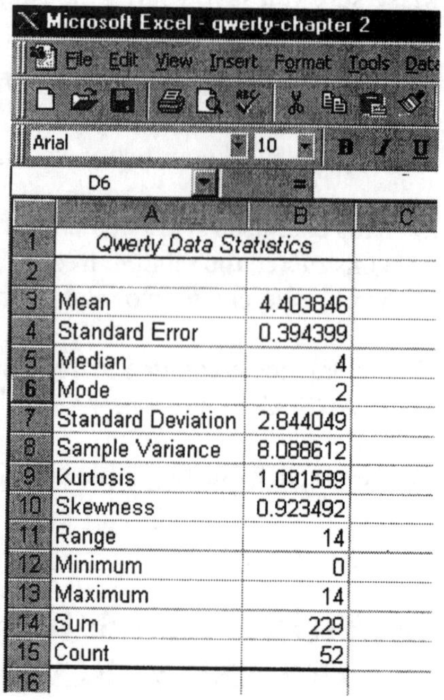

Microsoft Excel - qwerty-chapter 2

File Edit View Insert Format Tools Data

Arial 10 B I U

D6 =

	A	B	C
1	Qwerty Data Statistics		
2			
3	Mean	4.403846	
4	Standard Error	0.394399	
5	Median	4	
6	Mode	2	
7	Standard Deviation	2.844049	
8	Sample Variance	8.088612	
9	Kurtosis	1.091589	
10	Skewness	0.923492	
11	Range	14	
12	Minimum	0	
13	Maximum	14	
14	Sum	229	
15	Count	52	
16			

11) To remove the bottom bar, select the two cells containing the bottom border. Click on the down arrow by the frame icon, and select the **No Border** icon in the displayed table.

Interpreting the Output in the Descriptive Output Table

Below is a brief description of each of the measures included in Descriptive Statistics.

- **Mean:** The arithmetic average of the numbers in your data set.

- **Standard Error:** This is computed by using the formula $S / \sqrt{n}$ where S is the sample standard deviation and n is the number of observations.

- **Median:** This is the data value that splits the distribution in half. To determine the value of the median, the observations are first arranged in either ascending or descending order. If the number of observations is even, the median is found by taking the arithmetic average of the two middle values. If the number of observations is odd, then the median is the middle observation.

- **Mode:** This is the observation value associated with the highest frequency. **Caution:** Three situations are possible regarding the mode: 1) If all values occur only once in a distribution, Excel will return #N/A. 2) If a variable has only one mode, Excel will return that value. 3) If a variable has more than one mode, Excel will still return only one value. The value used will be the one associated with the modal value that occurs first in the data set. To check the accuracy of the mode, it would be wise to create a frequency distribution.

- **Standard Deviation:** This is computed using the formula: $S = \sqrt{\frac{\sum (X - \overline{X})^2}{n-1}}$

- **Sample Variance:** This is the standard deviation squared.

- **Kurtosis:** This number describes a distribution with respect to its flatness of peakedness as compared to a normal distribution. A negative value characterizes a relatively flat distribution. A positive value characterizes a relatively peaked distribution.

- **Skewness:** This number characterizes the asymmetry of a distribution. Negative skew indicates that the longer tail extends in the direction of low values in the distribution. Positive skew indicates that the longer tail extends in the direction of the high values.

- **Range:** The minimum value is subtracted from the maximum value.

- **Minimum:** The lowest value occurring in the data set.

- **Maximum:** The highest value occurring in the data set.

- **Sum:** The sum of the values in the data set.

- **Count:** The number of values in the data set.

CREATING PARTICULAR SAMPLE STATISTICS USING THE FUNCTION WIZARD

If you just want to know particular values, without producing the entire table of Descriptive Statistics, you can use the **Function** wizard, and select just the options that you want to use.

1) Decide which values you want to compute. We will compute the Mean, Count, Maximum, Minimum and Standard Deviation. Copy your data on Qwerty Keyboard Ratings into column A of a new worksheet.

2) In cells C1 through C5, type in the words "Mean", "Count", "Maximum", "Minimum" and "Standard Deviation".

3) Move to cell D1, and click on **Insert** from the main menu bar, then select **Function**, or click on the **Function** icon.

4) Click on **Statistical** from the **Function Category**, and then select **Average** under **Function Name**.

5) In the input box by **Number 1**, type in "A2:A53", and click on **OK**. You should see the average for your data returned in cell D1.

6) Move to cell D2, and from the **Function** menu, click on **Count.** Again, type in "A2:A53" in the input box and click on **OK**.

7) In cells D3, D4 and D5, repeat this process, clicking on **Min, Max**, and **STDEV** from the **Function** menu.

8) You should now have the information below:

Mean	4.403846
Count	52
Minimum	0
Maximum	14
Standard Deviation	2.844049

TO PRACTICE THESE SKILLS

You can apply the technology skills covered in this section by working through the following problems. Make sure you save your work with a file name that is indicative of the material covered.

1) Load the data from the CD for Data Set 2 in Appendix B (TEXTBOOK.XLS) into a new worksheet. Use this data to complete exercise 13 on page 67 of your text. Load the data from the CD for Data Set 17 in Appendix B (BOSTRAIN.XLS) into a new worksheet. Use this data to complete exercise 15 on page 67 of your text. Make sure you save your work with a file name that is indicative of the material covered.

2) Use one of the data sets that you previously worked with above, and complete exercise 21 on page 68 of your text. You should pick an actual value for *k*, and create new columns where each value in your original data has had *k* added, and each original value has been multiplied by *k*. Compute the mean, median, mode and midrange for your original data, as well as each of these new columns, and compare your results.

3) Load the data from the CD for Data Set 7 from Appendix B (BEARS.XLS). Complete 23 on page 68 of your text. You will need to arrange the data in order before you delete any values from the top and bottom of the list.

SECTION 2-6: MEASURES OF POSITION

Z-Scores

When looking at a set of data, it is often useful to know how far a particular score falls from its mean. We can measure the position of a particular value with respect to the mean using z-scores. We know that if a value is more than 2 standard deviations away from the mean of the data set, it can be considered "unusual". Remember that whenever a value is below the mean, the corresponding z-score will be negative. We will create a column of z scores for the data on Qwerty Keyboard Ratings.

1) Use the worksheet where you created the Mean, Count, Minimum, Maximum and Standard Deviation, or create a new worksheet with the Qwerty Keyboard Ratings and compute the Mean and Standard Deviation according to the directions in section 2–5.

2) If your original data is in column A, position your cursor in cell B1, and type "Standard Score".

3) Position your cursor in cell B2, and from the **Function** menu (accessed either through **Insert, Function**, or by pressing the function icon), click on **Statistical** and **Standardized.**

4) In the dialog box, type in "A2" for where your **x** value is found. In the box for the **Mean**, type in "$D$1". Notice this is an absolute address, which will remain constant for all computations, whereas the x values will be updated when you use the fill command. In the box for the **Standard_dev**, type in "$D$5".

5) Click on **OK**. You will now see the Standardized score for the first number in your data list (2) in cell B2. (The completed **STANDARDIZE** dialog box is shown on the next page.)

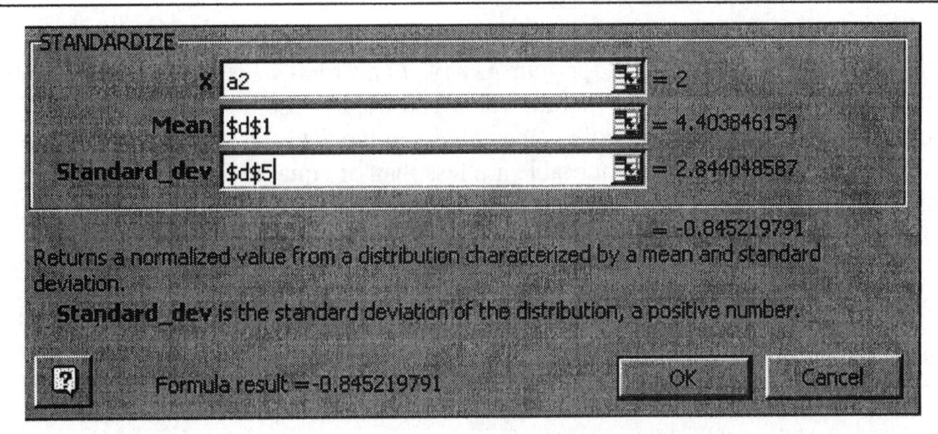

6) Since z-scores are normally reported to only two decimal places, you should format this column to show only two decimal places. With your cursor in cell B2, click on **Format** from the main menu. Click on **Cells**, click on **Number**, and then indicate 2 in the box by Decimal Places. Click on **OK.** You should now see the value –0.85 displayed. This indicates that the value 2 is approximately .85 standard deviations below the mean.

7) Use the fill handle to copy the formula down for the rest of the values in your data set. Notice that your values have z-scores ranging from –1.55 (indicating that 0 is 1.55 standard deviations below the mean) to 3.37 (indicating that 14 is 3.37 standard deviations above the mean).

Measures of Position

We can use Excel to find the three Quartiles, ten Deciles or 99 Percentiles for a data set. There are two functions within Excel that allow us to do this quickly: **Quartiles** and **Percentile**. There is no function for Deciles, but by recognizing that Deciles are the 10th, 20th, 30th, etc. percentiles, we can easily use the Percentile function to create the Decile values.

Quartiles

1) Using the data on Qwerty Keyboard Ratings, copy the data to column A of a new worksheet.

2) Suppose we want to find the first, second and third quartiles for the data. In cell C1, type in "Quartile". In cells C2 through C4, type in 1, 2, 3.

3) Move your cursor to cell D2 and click on the **Function** icon on your menu bar, or click on **Insert, Function.**

4) Under **Function Category**, click on **Statistical.** Under **Function name**, click on **Quartile.**

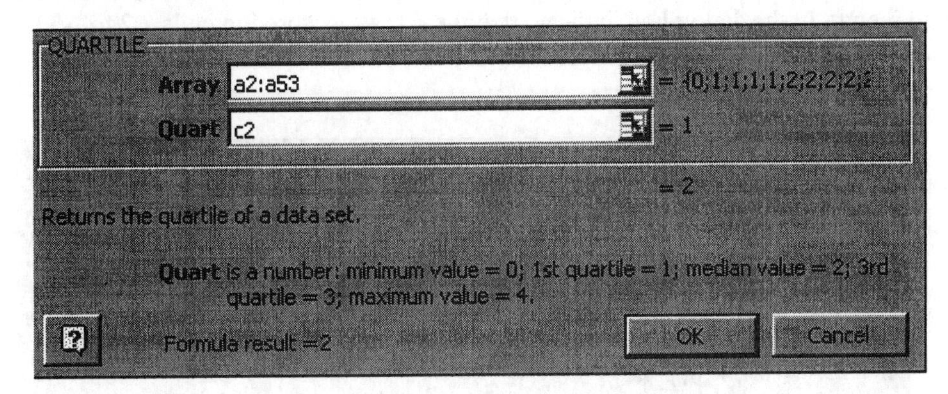

5) In the dialog box, type in "A2:A53" for your **Array**. (This represents the data values in your set.) In the **Quart** box, type in "C2" to indicate that the value you want to use for k is in cell C2. Then click on **OK**. You should see that the second quartile is the value 2 in the data set. Roughly speaking, this means that about 25% of the sorted values in your table are less than or equal to 2. We can say that at least 25% of the sorted values will be less than or equal to 2 and at least 75% of the sorted values will be greater than or equal to 2.

6) Use the fill handle to copy the formula down into the next two cells. You should see the values below:

Quartile	Value
1	2
2	4
3	6

7) To see that these values make sense in terms of your data, select column A by positioning your cursor on the A at the top of the column, and clicking once. Your entire column should now be selected. Then click on the **Sort** icon which shows from A to Z in your menu bar. This means that your data will be sorted from smallest to largest value.

Note:
Notice that there are 20 values which are less than or equal to 2. Since 20/52 represents approximately 38%, we can see that at least 25% of the data values are less than or equal to 2. The reason why we do not get a ratio closer to 25% is because 2 appears so many times in the data.

There are 27 values which are less than or equal to 4. Since 27/52 represents approximately 52%, we can see that at least 50% of the data values are less than or equal to 4.

There are 40 values which are less than or equal to 6. Since 40/52 represents approximately 77%, we can see that at least 75% of the data values are less than or equal to 6.

Percentiles

Let's now calculate the 10th, 20th, 30th,....90th percentiles. These values will correspond to the 1st, 2nd, 3rd,....9th deciles.

1) Using our previous worksheet where we computed the quartiles, position your cursor in cell F1 and type in "Percentile". Starting in cell F2, enter the values .1, .2, .3, .4,...., .9.

2) Move to cell G2, and click on the **Function** icon on the main tool bar. Click on **Percentile** from the **Statistical** menu. In the dialog box, indicate that the **Array** is stored in cells A2 to A53. Indicate that the k value (for the kth percentile) is stored in cell F2. Click on **OK.**

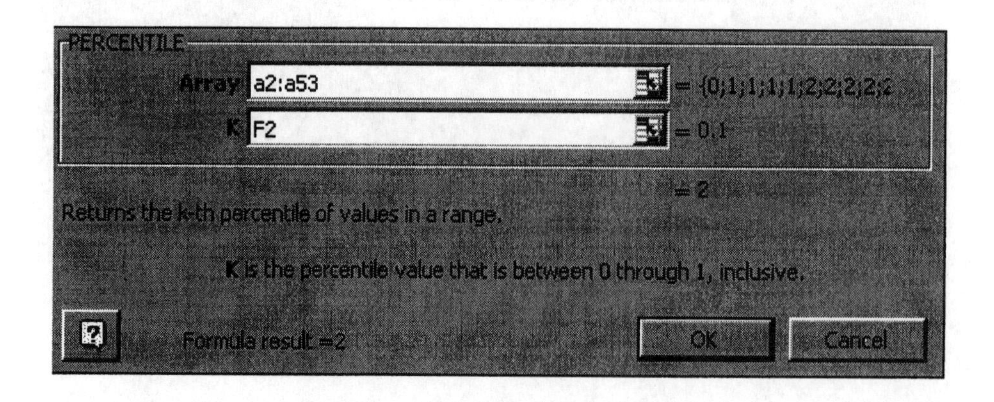

3) You should now see the value 2 in cell G2. Use the fill handle to copy the formula down the rest of the column. You should see the following table:

Percentile	
0.1	2
0.2	2
0.3	2
0.4	3
0.5	5
0.6	5
0.7	6
0.8	7
0.9	8

4) Remember that at least k % of the data values are less than or equal to these values.

TO PRACTICE THESE SKILLS

You can apply the technology skills covered in this section by working through the following exercises.

1) Load Data Set 1 in Appendix B from your CD into a new worksheet (COLA.XLS). Using the weights for Coke, find the z scores for this data. Are any of the weights of Coke unusual? Why or why not?

2) Using the weights of regular Coke, find Q1, Q2, and Q3. What is the true percentage of values that are below each of these values?

3) Using the weights of regular Coke, find the 10th through 90th percentiles. Determine the true percentage of values that are below the value representing your 80th percentile.

SECTION 2-7: EXPLORATORY DATA ANALYSIS

BOXPLOTS

Excel is not designed to generate boxplots. You can use the **DDXL** Add-In that is supplied with your book to generate this type of graph.

1) Open a worksheet where you have the data on Qwerty Keyboard Ratings entered in column A. Make note of the range where your actual values are stored. If you entered the name Qwerty in cell A1, and the data directly below this, your values will be contained in cells A2:A53.

2) Click on the **DDXL** command on the main menu bar.

3) Select **Charts and Plots**. Under the **Function Type**, select the option **Boxplot**.

4) Click on the **pencil** icon at the bottom of the screen on the left hand side. Type in the cells where your data is stored in your Excel worksheet. In this case, you should type in "A2:A53".

5) Click on **OK** in the bottom right hand side of the screen. You will be taken to the **DDXL** screen which should appear as the one displayed below. Notice that the boxplot is in the upper left hand corner, and that the Summary Statistics box can be found directly below that.

6) Click on the triangle in the upper left corner of the **Boxplot** screen, and select **Plot Scale**. Change the settings for the Y axis as shown in the **Scale Plot** window. Click on **OK**.

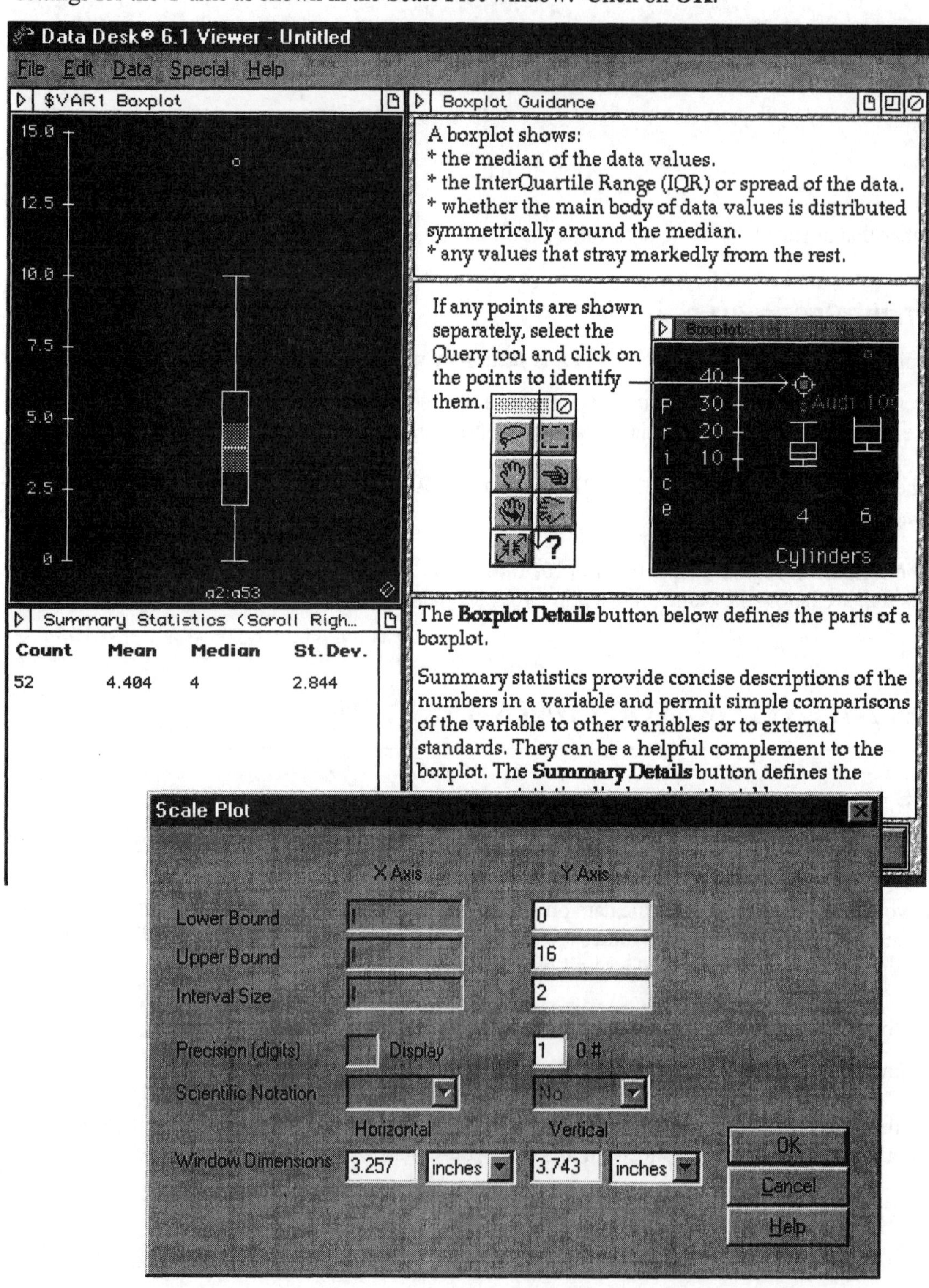

8) Click and hold on the diamond shape in the lower right hand corner of the **Boxplot** screen, and drag this corner out to create a larger graph. Your graph should look like the one below.

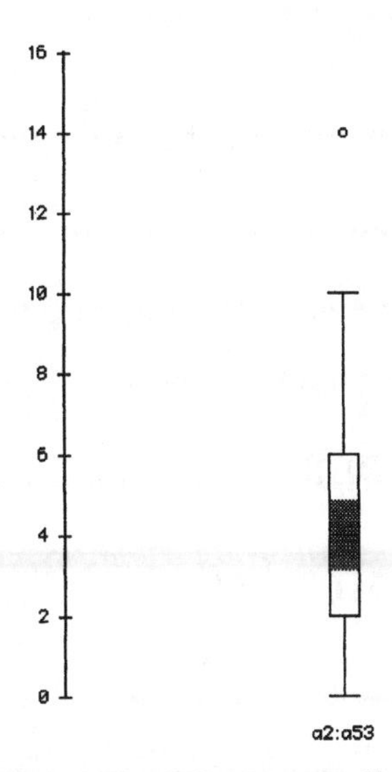

9) Click anywhere in the **Summary Statistics** window to activate that screen. Again, click and hold on the diamond in the lower right hand corner of this box, and drag to the right to expand the amount of the box that can be seen. You want to be able to see all the scores shown below.

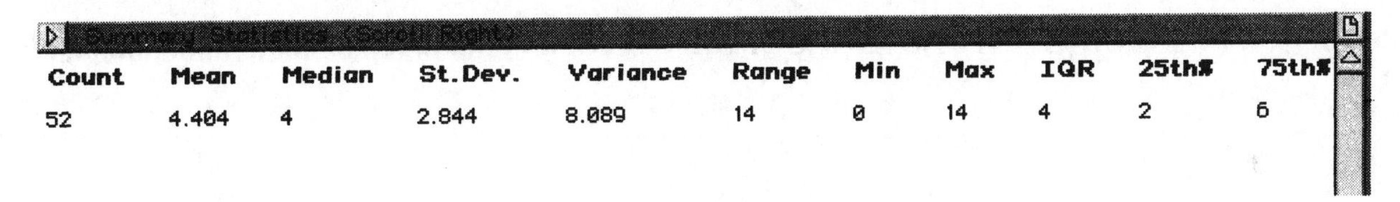
Summary Statistics (Scroll Right)

Count	Mean	Median	St.Dev.	Variance	Range	Min	Max	IQR	25th%	75th%
52	4.404	4	2.844	8.089	14	0	14	4	2	6

TO PRACTICE THESE SKILLS

You can apply the technology skills covered in this section by working through the exercises on page 102 of your textbook. Remember that you may have already loaded some data from the CD into an Excel workbook for work from a previous section. You can open this file, and create a new worksheet within the file for any additional work you do with this particular data set.

CHAPTER 3: PROBABILITY

SECTION 3-1: OVERVIEW

This chapter in your textbook covers the basic definitions and concepts of probability. While many of those concepts are straightforward and can be done without the use of technology there are some features of Excel that can be utilized while working through the material found in this chapter. The following list contains an overview of the topics and functions that will be introduced within this chapter.

PIVOT TABLE
This feature is used to generate a worksheet table that summarizes data from a data list. This allows you to obtain category counts.

RANDBETWEEN
This function creates columns of random numbers that fall between the numbers you specify. A new random number is returned every time the worksheet is calculated. The function appears in the following format:
RANDBETWEEN (bottom, top) where bottom is the smallest integer RANDBETWEEN will return and top is the largest integer RANDBETWEEN will return.

FACT
Returns the factorial of a number. The factorial of a number is equal to 1•2•3•...• number.
FACT(number) where number refers to the nonnegative number you want the factorial of. If number is not an integer, it is truncated.

PERMUT
Returns the number of permutations for a given number of objects that can be selected from a larger group of objects. A permutation is any set or subset of objects or events where internal order is significant.
PERMUT(number, number_chosen) where number is an integer that describes the number of objects and number_chosen is an integer that describes the number of objects in each permutation.

COMBIN
Returns the number of combinations for a given number of items. Use COMBIN to determine the total possible number of groups for a given number of items.
COMBIN(number, number_chosen) where number is the number of items and number_chosen is the number of items in each combination.

SECTION 3-2: PIVOT TABLES

In Excel it is possible to generate a table, called a **pivot table,** that can be used to summarize both quantitative and qualitative variables contained within a database. A **pivot table** provides us with a mechanism for creating subgroups (or samples), and gives us the ability to find sums, counts, averages, standard deviation and variance of both a sample and population.
As you look at the table presented in the Chapter problem at the start of Chapter 3 keep in mind that this information had to be tabulated and cross referenced in order to present the summary you see there. Today, with the use of technology, this could be done by means of a pivot table.

CREATING A PIVOT TABLE:

We will create a pivot table using data found on the CD-ROM data disk that accompanies your textbook. Begin by opening Excel. Open the data file "CARS".

Excel provides a **PivotTable Wizard** similar to the ChartWizard you have used to generate tables and graphs.

To create a Pivot Table:

1) On the menu bar click on **Data**, highlight **Pivot Table Report** and click.

The Pivot Table Wizard – Step 1 of 4 dialog box will appear. Make sure that **Microsoft Excel list or database** is selected. Then click on **Next**.

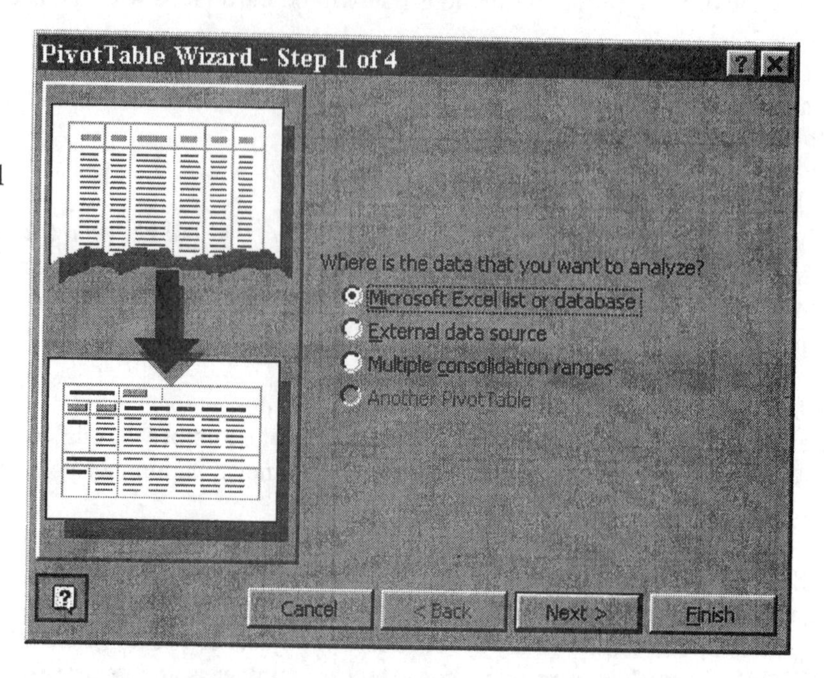

2) Indicate the worksheet range that contains the data you want to use. Drag the cursor across from cell A1 to cell E30. In the **Range window** you should see Sheet1!$A$1:$E$30. Click on **Next**.

3) The third dialog box, shown below, presents options for setting up the Pivot Table. The dialog box displays a graphical layout of an empty pivot box as well as field buttons that correspond to the columns in your Excel worksheet. You can specify the layout of the table by dragging fields into any of the four areas: COLUMN, ROW, DATA and PAGE. You can drag as may buttons as you like. If you make a mistake when dragging buttons, you can drag them off and rearrange the Pivot Table as necessary.

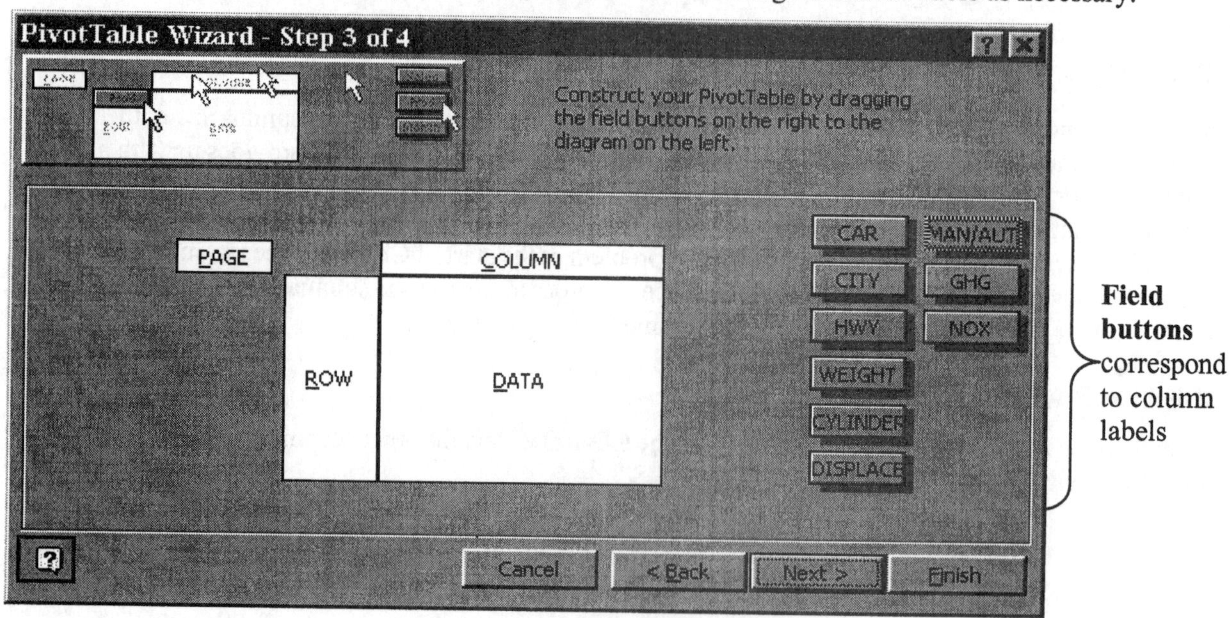

Field buttons correspond to column labels

4) To help you create your first Pivot Table

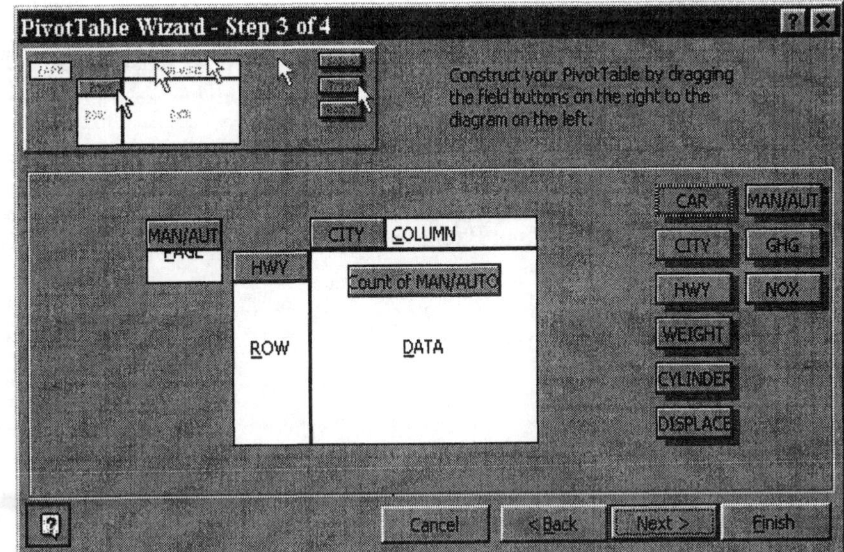

   a) Drag MAN/AUT to **Page**
   b) Drag CITY to **Column**
   c) Drag HWY to **Row**
   d) Drag MAN/AUT to **Data**

Notice that when we move Man/Auto to Data it now reads "Count of Man/Auto".

The table you create will display the number of city and highway miles broken down by the type of transmission (manual or automatic).

5) Click on **Next** to move to go to the 4th dialog box.

6) Select where the Pivot Table is to be displayed – either as part of the current worksheet or in a separate worksheet. Click on **Options** for a list of choices to be included in the display of your Pivot table. Then click on **Finish**. The completed table is shown below.

K	L	M	N	O	P	Q	R	S	T	U
MAN/AUTO	(All)									
Count of MAN/AUTO	CITY									
HWY	17	18	19	20	22	23	28	29	32	Grand Total
24	2									2
26	2									2
27			3							3
28		1								1
29				2						2
30			1		1	1				3
31						1				1
32						2				2
33					1					1
34								1		1
37							1		1	2
Grand Total	4	1	4	2	2	4	1	1	1	20

7) Click on the **down arrow** by MAN/AUTO to see the data summary for only manual transmissions or only automatic transmissions.

Note:
Once you have created a Pivot table you can refine it. Right click on any cell within the Pivot Table for a list of options. The Wizard option will take you back to the Pivot Table Wizard used to create your existing table. Double click on the **field button** within the Pivot Table to open the Pivot Table Field dialog box for options concerning fields. As always your authors encourage you to explore and experiment with the different options available to you.

PROBABILITIES

The pivot table shown on the right was created to display the number of highway miles and its relationship to the type of car transmission (automatic or manual). Using a pivot table allows you to determine probability information more easily than sorting through the information presented to you in the original Excel worksheet.

Count of MAN/AUTO	MAN/AUTO		
HWY	A	M	Grand Total
24	2		2
26	2		2
27	3		3
28		1	1
29	1	1	2
30	2	1	3
31	1		1
32	1	1	2
33		1	1
34	1		1
37	1	1	2
Grand Total	14	6	20

TO PRACTICE THESE SKILLS

Use the information presented in the following Excel worksheet to create a Pivot Table that will display the total number of credits broken down by Religion and Major for

a) male and female students combined
b) only female students
c) only male students

	A	B	C	D	E	F
1	**SEX**	**AGE**	**MAJOR**	**CREDITS**	**GPA**	**RELIGION**
2	M	22	Liberal Arts	19	2.5	Jewish
3	M	23	Computer Science	14	3.7	Protestant
4	F	19	Criminal Justice	17	3.8	Protestant
5	M	22	Mathematics	18	2.4	Protestant
6	F	21	English	13	2.5	Jewish
7	F	23	Liberal Arts	18	3	Catholic
8	F	22	Liberal Arts	17	3.2	Catholic
9	M	22	Liberal Arts	18	3.6	Protestant
10	M	22	Education	13	3.5	Catholic
11	F	21	English	17	2.7	Protestant
12	F	22	Criminal Justice	15	2.5	Catholic
13	M	23	Computer Science	15	3.9	Jewish
14	F	22	Computer Science	17	3.3	Jewish
15	M	21	Engineering	12	2.5	Protestant
16	F	22	Engineering	18	4	Protestant
17	F	21	Mathematics	16	3.6	Catholic
18	F	22	Liberal Arts	18	3.4	Jewish
19	M	19	English	17	2.9	Jewish
20	M	21	Criminal Justice	17	3	Catholic
21	F	20	Engineering	16	2.8	Catholic

SECTION 3-3: GENERATING RANDOM NUMBERS

Begin by making cell A1 your active cell.

1) To generate a random set of numbers in Excel, click on **Insert**, highlight **Function** and click. The following **Paste Function Dialog box** will open. We have already made use of Excel's built in Statistical functions in previous sections.

   a) To generate random numbers highlight **Math & Trig** as seen on the right.
   b) Scroll through the list of function names and highlight **RANDBETWEEN.**
   c) Click **OK.**

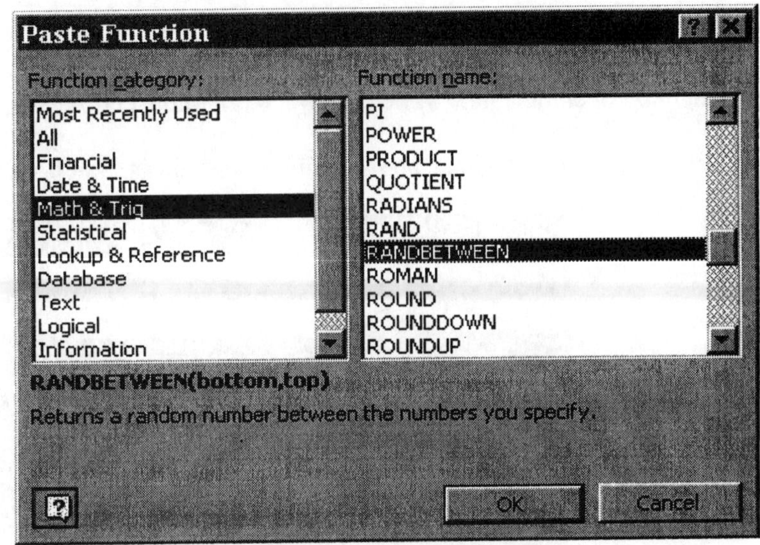

2) The dialog box, shown on the next page, should now be located in the upper left-hand corner of your Excel worksheet. Fill in the lowest and highest values between which the random number will fall. For example, if you wish to generate a list of possible values that can occur when you roll a single die then the bottom number would be 1 and the top value would be 6. Click **OK.**

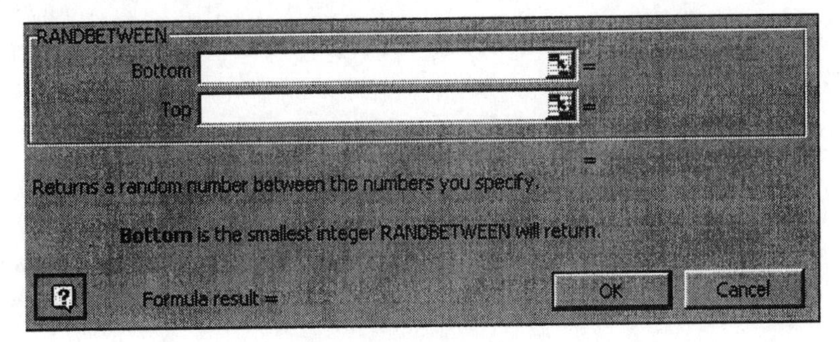

A random number between 1 and 6 should now appear in cell A1. To generate 25 such numbers copy (by dragging) this cell entry to cell A25. When you have completed this task you should have 25 entries in the first column, each between 1 and 6. A new random number is returned every time the worksheet is calculated.

SECTION 3-4: PROBABILITIES THROUGH SIMULATION

The goal of every statistical study is to collect data and to use that data to make a decision. Often collecting data or repeating a trial a large number of times can be impractical. With the use of technology we can often simulate an event.

Using the definition from the textbook:

"A simulation of an experiment is a process that behaves in the same ways as the experiment itself thus producing similar results."

The random number generator can be used to simulate a variety of statistical problems.

CREATING A SIMULATION

Let's consider the **Gender Selection problem** presented on page 151 at the start of section 3-6.

1) Begin by opening a new worksheet in Excel. Make cell A1 your active cell.

2) Using the method outlined above simulate 100 births. Let 0 = male and 1 = female.

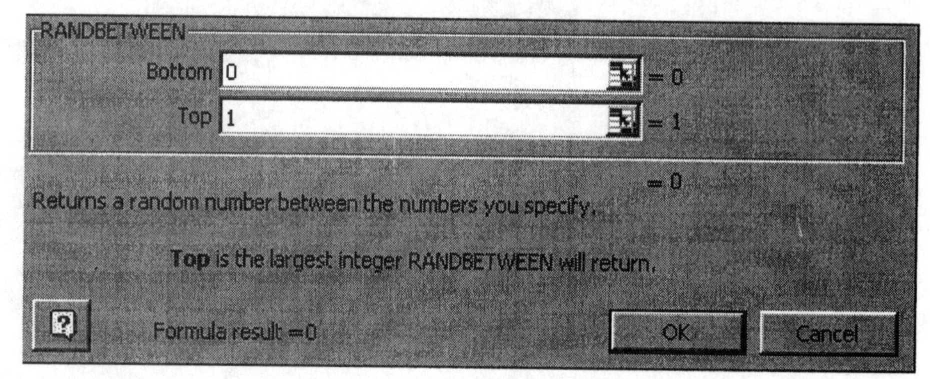

3) You can copy the information through to cell A100 or you can copy it through to cell A25. The by dragging, copy the information into columns B, C and D. The advantage of this second method is that you can see the 100 entries in your simulation. You may notice that the entry in your first cell change. Don't worry about this. The cells will continue to change until we do our statistical analysis.

4) To determine the probability that the newborn if female, count the number of 1's in your data and divide by the total number of entries.

It might be interesting to repeat the simulation several times and compare your results.

Simulating Multiple Events

Consider a problem that requires more than one event to occur such as the probability of a specific sum when two dice are tossed as mentioned in the example **Simulating Dice** found on the bottom of page 153 in section 3-6 of your textbook.

For the purpose of this example we wish to find the P(sum of 4) when two dice are tossed 50 times. One possibility would be to toss a pair of dice 50 times and record the sums after each roll. Using technology to simulate this experiment we will produce similar results.

1) Begin in a new worksheet with cell A1 as the active cell. Type "First Die" in cell A1, type "Second Die" in cell B1 and type "Sum" in cell C1.

2) Begin in cell A2. Generate a *column* of 50 random values between 1 and 6.

3) Repeat the process beginning in cell B2. As before, you may notice that the cell entries in your first column change. This is normal. The cells will continue to change until we do our statistical analysis.

4) In cell C2 type =Sum(A2 + B2) and press **Enter**.

5) Copy this formula down through to cell C51. This should give you the sums of the toss of two dice. Once again you will notice that the entries in your first two columns have changed. This is normal. You should be able to see rather easily that the sum of the first two column entries is reflected in the third column.

6) At this point it is possible to determine how many of our simulated tosses of the dice yield a sum of 4.

TO PRACTICE THESE SKILLS

You can practice the skills learned in Sections 3-3 and 3-4 by working through the following problem.

Some role-playing games use dice that contain more sides than the traditional six sided dice most of us are familiar with. Assume we are playing such a game and that we are using a pair of ten sided dice which contain the numbers one through ten each die.

a) Use the Random Number Generator to simulate rolling a pair of ten sided dice fifty times.
b) Use the results to determine a list of possible sums from the fifty rolls of the dice.
c) Find the probability of rolling a sum of 20.

SECTION 3-5: FACTORIAL

In looking at the **Survey Question** example found on page 157 in section 3-7 of the textbook we found that we were required to multiply $5 \cdot 4 \cdot 3 \cdot 2 \cdot 1$. This product can be represented by 5! which is read as "five factorial". What appears to be an exclamation point after the 5 is really a **factorial symbol (!).** The factorial symbol indicates that we are to find the product of decreasing positive integers.

In Excel we can locate the **factorial function** using the same method we did for generating a random number.

1) Click on **Insert**, highlight **Function** and click. The **Paste Function Dialog box** will open.

2) Highlight **Math & Trig.**

3) Scroll through the function names and highlight **FACT.**

4) Click **OK.**

5) In the **FACT** dialog box enter that number you wish to expand by using factorials. **FACT(5)** equals $5 \cdot 4 \cdot 3 \cdot 2 \cdot 1 = 120$.

6) Click **OK.**

SECTION 3-6: PERMUTATIONS AND COMBINATIONS:

Problems involving permutations and combinations such as those found in section 3-7 of your textbook can be done fairly quickly and quite easily with Excel. A permutation is an ordering or arrangement of any set or subset of objects or events where order is significant. Permutations are different from combinations, for which the order is not significant.

The **PERMUT (permutations)** function can be found in the Function Dialog box in the **Statistical** function category.

The **COMBIN (combinations)** function can be found in the Function Dialog box in the **Math and Trig** function category.

Consider the **Frank Sinatra example** found on page 160 in your textbook. Use the **PERMUT function dialog** box:

Number – an integer that refers to the total number of objects - in this case top 10 songs.

Number_chosen - an integer that identifies the number of objects in each permutation. – in this case 3.

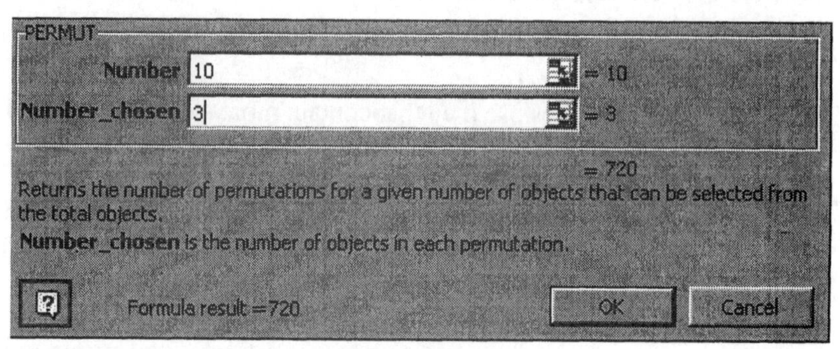

The result is 720 different possible arrangements of 3 songs taken from the top 10.

Consider the **Elected Offices example** found on the bottom of page 161 in your textbook. Note that we use combinations for part (a) of this problem because order does not count. Use the **COMBIN function dialog** box:

Number - the number of items – in this case 9 members.

Number_chosen - the number of items in each combination – in this case 3 person committee

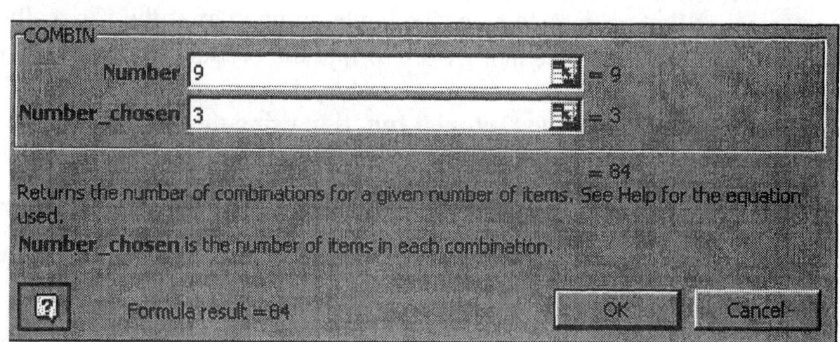

There are 84 different possible committee of 3 board members.

CHAPTER 4: PROBABILITY DISTRIBUTIONS

SECTION 4-1: OVERVIEW

Excel has functions built into it that can be used to calculate the probabilities associated with several different probability distributions. Computing these probabilities by hand can be very time consuming. Although tables are available for some distributions, these are also limited in scope. Excel provides you with a tremendous amount of flexibility in creating these distributions quickly and efficiently. The following list contains an overview of the new functions that will be introduced within this chapter.

FILL SERIES
This feature enables us to quickly and efficiently enter a string of consecutive numbers in a column or row.

BINOMDIST
This function returns the individual term binomial distribution probability.

POISSON
This function returns the Poisson probability that a particular number of occurrences of an event will occur over some interval.

SECTION 4-2: RANDOM VARIABLES AND PROBABILITY DISTRIBUTIONS

We will consider the probability distribution given in Table 4-1 on page 181 of your book. We will work with the values given, and create a probability histogram, as well as consider how we can use Excel to compute the mean and standard deviation of this probability distribution.

1) Type "x" in cell A1 of a new worksheet, and "P(x)" in cell B1. Then enter the values given in table 4-1 starting in cells A2 and B2 respectively.

2) Click on the **Chart Icon**, or click on **Insert** in the menu bar, then click on **Chart.**

3) Click on **Column** under **Chart Type** and the first option under **Chart sub-type**. Then press **Next**.

4) With your cursor positioned in the dialog box by **Data Range**, select cells A2 through A16. Notice that the box comes up showing: =Sheet1!$A$2:$A$16.

5) Make sure that the bubble in front of **Columns** is selected.

6) Click on the **Series** tab, and position your cursor in the dialog box by the word **Values** . Select cells B2 through B16 in your worksheet. Notice that the box will show: =Sheet1!$B$2:$B$16.

7) Click on **Next**, and give your graph and your axes appropriate names. Then click on **Finish**.

8) You will need to modify your graph. You can click on the Legend box, and delete this. You should double click on one of the bars, and in the **Format Data Series** dialog box, click on **Options**, and set your gap width to 0. You will probably also want to resize your chart area. If you need more specific instructions, look back in Chapter 2, and follow the appropriate instructions in section 2-3.

9) When you are done, your probability distribution should look similar to the one below.

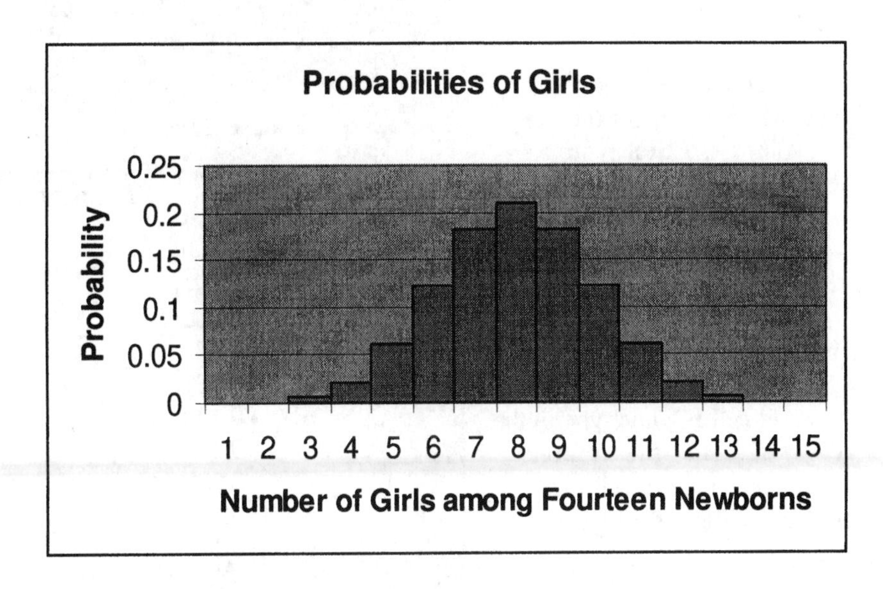

MEAN, VARIANCE AND STANDARD DEVIATION

We can compute the mean, variance and standard deviation of a probability distribution according to formulas 4 –1, 4- 3, and 4-4 given on pages 184 and 185 of your book. To use Excel to work with these formulas, follow the steps below.

1) In cell D1, type in "x * P(x)".

2) In cell D2, type in the following formula: = A2 * B2. This tells Excel to multiply the random variable in cell A2 by its associated probability in cell B2. After pressing **Enter**, you should see a 0 in cell D2, since both values being multiplied are 0.

3) Position your cursor back in cell D2. Then use the fill handle to copy the formula down through cell D16.

4) Formula 4-1 tells us that we need to add these products up. Position your cursor in cell C17, and type "Sum".

5) Position your cursor in cell D17, and click on the **Summation** icon in the tool bar. Notice that the formula: =SUM(D2:D16) appears in cell D17, and the column of numbers directly above this cell is selected. Press **Enter.** You should now see the value 6.993 in cell D 17. This is the mean of the probability distribution.

6) Move to cell C19, and type the word "Mean". In cell C20, type "Variance", and in cell C21, type in "Std.Dev".

7) Move to cell D19 and enter the formula: =d17. Notice that this is an absolute address. When you press **Enter**, you should see the value for the mean in cell D19.

8) To compute the variance, we will use Formula 4-3. Position your cursor in cell E1, and type in the formula: + x^2*P(x).

9) Move your cursor to cell E2, and type in the formula: = A2^2*B2. Then press **Enter**. This formula takes the value in cell A2, squares it, and then multiplies it by the value in cell B2.

10) Reposition your cursor in cell E2, and use the fill handle to fill the column down through E16.

11) Position your cursor in cell E 17, and click on the summation icon in the tool bar. You should see the formula: =SUM(E2:E16) appear in cell E17. Press **Enter**, and you will see the value 52.467.

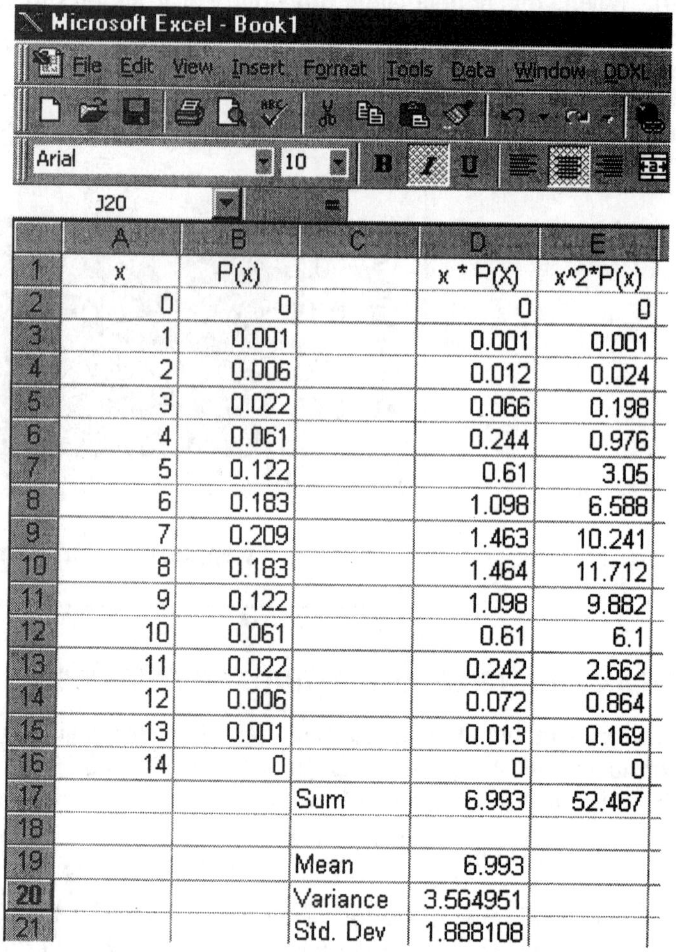

Microsoft Excel - Book1

File Edit View Insert Format Tools Data Window DDXL

Arial 10

J20

	A	B	C	D	E
1	x	P(x)		x * P(X)	x^2*P(x)
2	0	0		0	0
3	1	0.001		0.001	0.001
4	2	0.006		0.012	0.024
5	3	0.022		0.066	0.198
6	4	0.061		0.244	0.976
7	5	0.122		0.61	3.05
8	6	0.183		1.098	6.588
9	7	0.209		1.463	10.241
10	8	0.183		1.464	11.712
11	9	0.122		1.098	9.882
12	10	0.061		0.61	6.1
13	11	0.022		0.242	2.662
14	12	0.006		0.072	0.864
15	13	0.001		0.013	0.169
16	14	0		0	0
17			Sum	6.993	52.467
18					
19			Mean	6.993	
20			Variance	3.564951	
21			Std. Dev	1.888108	

12) Position your cursor in cell D20 and type in the formula: =$E$17-$D$17^2. This will take the sum of your products and subtract the square of the mean from this sum. This corresponds to Formula 4-3 found on page 185 in your book.

13) Now position your cursor in cell D21, and type in the following formula: =SQRT($D$20). This will take the square root of the variance, which will produce your standard deviation.

EXPECTED VALUE

Notice that the value that you computed in your worksheet for the summation of the products of your random variables and their corresponding probabilities can also be called the expected value of a discrete random variable.

TO PRACTICE THESE SKILLS

You can apply these technology skills by working on the following exercises. Make sure you save your work using a file name that is indicative of the material contained in your worksheets.

1) Enter the data from exercise 5 on page 191 of your textbook. Use Excel, and the appropriate formulas to find the mean and standard deviation for this data.

2) Copy the work that you did for exercise 1 into another worksheet. Replace your P(x) values with the data in exercise 8 on page 191 of your textbook. The other values in your worksheet should be automatically updated to reflect this new information.

3) Use your data from exercise 1 to complete the following:

   a) Add 3 to each x value. What is the relationship between your new mean and standard deviation compared to the values from the original data?

   b) Multiply each x value by 2. What is the relationship between your new mean and standard deviation compared to the values from the original data?

   c) Multiply each x value by 3, and then increase by 5. What is the relationship between your new mean and standard deviation compared to the values from the original data?

SECTION 4-3: BINOMIAL PROBABILITY DISTRIBUTIONS

Binomial variables take on only two values. One of these values is generally designated as a "success" and the other a "failure". We typically see the probability of a "success" denoted by the letter p and the probability of failure denoted by the letter q. The sum of p and q must equal one, since success or failure are the only possible outcomes.

Suppose we consider a binomial distribution where $p = .65$ and there are 15 trials. Since there are 15 trials, we know that the random variable can take on the values between 0 and 15 inclusive.

To enter the column of random variables into Excel:

1) Enter a title for the column representing the random variable in cell A1. (If we knew what the random variable represented, we should use a suitable name, otherwise, we will use "x".)

2) In cell A2, type in the value 0, since this is the first value of the random variable. Press **Enter**.

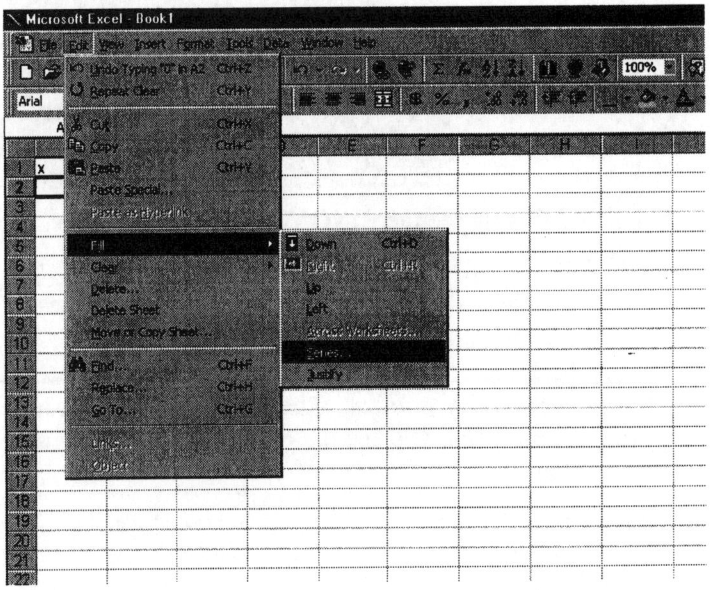

3) Reposition your cursor in cell A2. From the command bar, click on **Edit, Fill,** and then click on **Series.**

4) You will see the **Series** dialog box open. (This box is shown on the next page.)

5) Make sure that the bubble in front of **Columns** is checked.

6) Make sure that **Linear** is selected under **Type**.

7) Make sure that the **Step value** is set at 1.

8) Type in 15 for the **Stop value**, since there are 15 trials in the experiment.

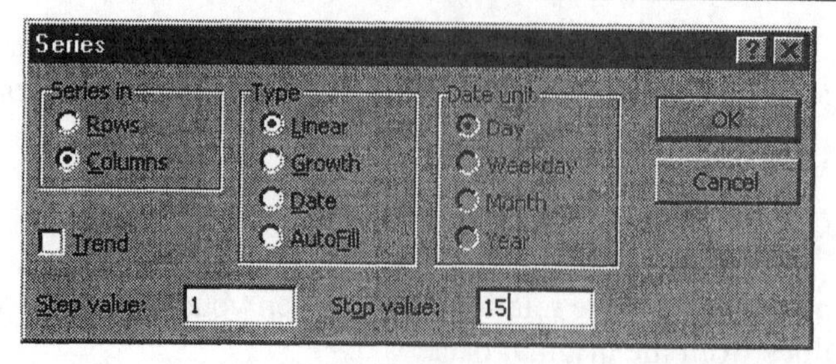

9) Click on **OK**. You should now see the whole numbers from 0 to 15 in column A.

10) Move to column B and type "P(x)" in cell B1. Press **Enter**. Your cursor should now be in cell B2.

11) In the command bar, click on **Insert**, and then click on **Function**. Notice the symbol preceding **Function** can also be found on the menu bar. You may activate this dialog box by clicking on this icon in the menu bar.

12) Under **Function Category**, click on **Statistical.** Under **Function name**, click on **BINOMDIST**. Then click on **OK**.

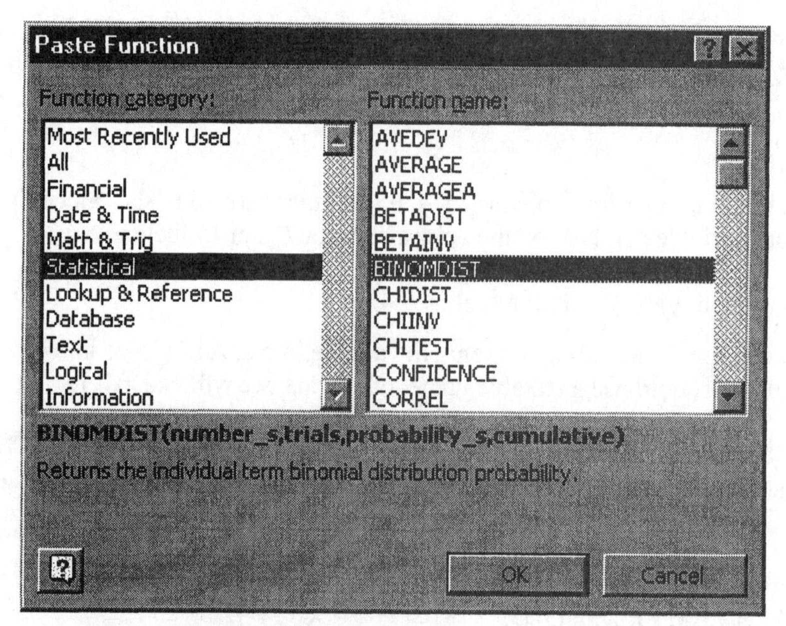

13) You will now see the **BINOMDIST** (Binomial Distribution) Input box. You need to complete the dialog box as follows:

   a) **Number_s** refers to the number of successes. You want to enter the cell address where this information is stored. Type in A2.
   b) **Trials** refers to the total number of trials. Type in 15.
   c) **Probability_s** refers to the probability of a success. For this experiment, type in the value .65.
   d) **Cumulative** will list the cumulative probabilities. Since we do not want these at this time, type in "False".

e) Click on **OK.**

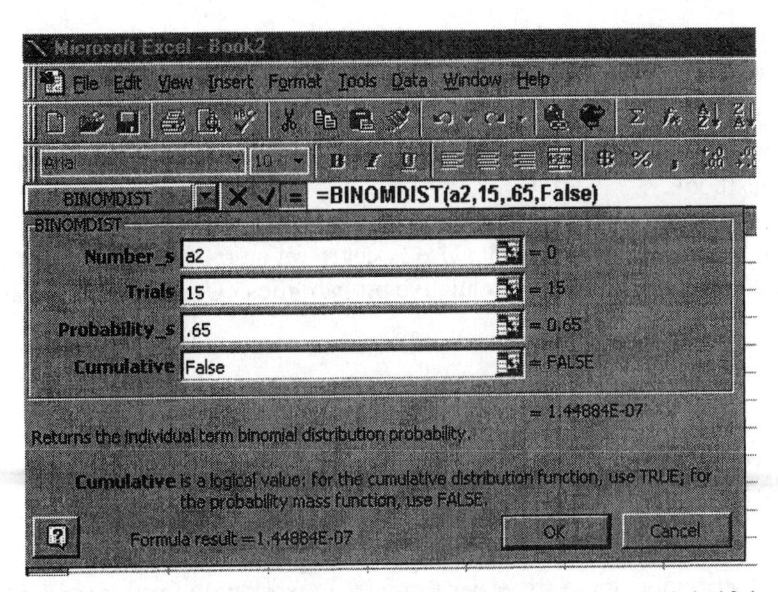

14) You will now see the probability of getting exactly 0 successes in 15 trials if the probability or success is .65. This value may be written in exponential notation. You can reformat the column to show the number with more decimal places.

15) With cell B2 activated, you can use the fill handle to fill in the remaining values in the table.

16) You can likewise create a column showing the cumulative probabilities. In your table, type "P(X<=x)" in cell C1. This represents the probability that X is less than or equal to x, where x refers to the corresponding random variable in column A.

17) Follow the steps above to access the **BINOMDIST** (Binomial Probability Distribution) Input box. The information entered will be the same, except that you will type in "True" in **cumulative.**

18) Again, use the fill handle to copy the function down through cell C17. Notice that cell C17 shows a value equivalent to 1. This should make sense to you, in that the sum of the probabilities in a probability distribution must add to 1.

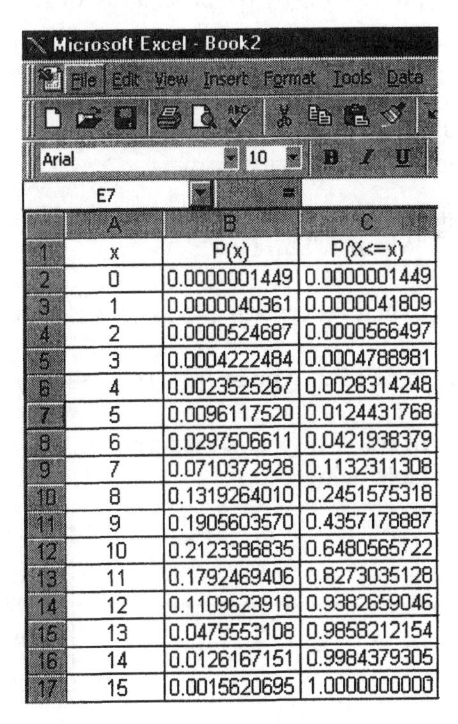

	A	B	C
1	x	P(x)	P(X<=x)
2	0	0.0000001449	0.0000001449
3	1	0.0000040361	0.0000041809
4	2	0.0000524687	0.0000566497
5	3	0.0004222484	0.0004788981
6	4	0.0023525267	0.0028314248
7	5	0.0096117520	0.0124431768
8	6	0.0297506611	0.0421938379
9	7	0.0710372928	0.1132311308
10	8	0.1319264010	0.2451575318
11	9	0.1905603570	0.4357178887
12	10	0.2123386835	0.6480565722
13	11	0.1792469406	0.8273035128
14	12	0.1109623918	0.9382659046
15	13	0.0475553108	0.9858212154
16	14	0.0126167151	0.9984379305
17	15	0.0015620695	1.0000000000

TO PRACTICE THESE SKILLS

You can apply these technology skills by working on the following exercises. Make sure you save your work using a file name that is indicative of the material contained in your worksheets.

1) Read exercise 33 on page 203 of your textbook. Use Excel to set up the probability distribution for this exercise. Create a column representing the cumulative probabilities. Use this column to answer the question asked in the textbook.

2) Read exercise 26 on page 202 of your textbook. Use Excel to set up the probability distribution for this exercise. Create a column representing the cumulative probabilities. Use your table to answer the questions asked.

SECTION 4-4: MEAN, VARIANCE, AND STANDARD DEVIATION FOR BINOMIAL DISTRIBUTION

Although you can use the formulas presented in section 4-2 to compute the mean, variance and standard deviation for the binomial distribution, there are easier formulas to work with for this particular distribution.

You can find the mean by multiplying the sample size (n) by the probability of success (p).

You can find the standard deviation by taking the square root of the product formed by multiplying the sample size (n) by the probability of success (p) and the probability of failure (q).

You should use formulas preceded by the "=" sign when you enter your information into Excel.

Suppose we used the values: n = 14, p = 0.5 and q = 0.5. You can follow the steps below to create a table for this binomial experiment.

1) In cells A1 through A5, type in: "Sample Size", "Probability of Success", "Probability of Failure", "Mean", "Std. Dev".

2) In cell B1 type in 14. In cells B2 and B3, type in 0.5.

3) In cell B3, type in =B1*B2. When you press **Enter**, you should see the value shown in the table below.

4) In cell B4, type in the formula: = sqrt(B1*B2*B3). Pressing **Enter** should return the value shown in the table below.

Sample Size	14
Probability of Success	0.5
Probability of Failure	0.5
Mean	7
Std. Dev	1.870829

TO PRACTICE THESE SKILLS

Once you have the table from above set up in a worksheet, you can change the values for sample size, probability of success and probability of failure. Your values for the mean and standard deviation should be automatically updated. You can copy this table, paste it in other cells of your worksheet, and then change the numbers in the copy of the table for a different problem.

1) Create a table similar to the one above, but using the data from exercise 5 on page 207 of your textbook. To complete part b, you want to compute the values that are 2 standard deviations above and below the mean. Add lines to your table which will use the numbers generated to compute the minimum usual value ($\mu - 2\sigma$) and the maximum usual value ($\mu + 2\sigma$).

2) Copy the table you created in exercise 1, and change the sample size and probabilities of success and failure to complete exercise 7 in your textbook.

3) Copy the table again, and change the sample size and probabilities of success and failure to complete exercise 11 in your textbook.

SECTION 4-5: CREATING A POISSON DISTRIBUTION

In a Poisson distribution, the random variable x is the number of occurrences of the event in an interval. The interval can be time, distance, area, volume, or some similar unit.

Using the example on **World War II Bombs** on page 211 of your text, we can generate a table showing the probabilities that a region was hit 0, 1, 2, 3, 4 or 5 times. In this example, the computed mean is 0.929.

1) In cell A1, type in "x".

2) In cell A2, type in 0, and press **Enter.**

3) Reposition your cursor in cell A2, and click on **Edit, Fill, Series**.

4) In the **Series** dialog box, click on **Columns**, **Linear**, and enter a step value of 1 and stop value of 5. Then click on **OK.**

5) In cell B1, type in "P(x)".

6) In cell B2, click on **Insert**, **Function**, or click on the function icon (*fx*)on the menu bar. Click on **Statistical** in the **Function Category** box, and **POISSON** in the **Function Name** box. Then click on **OK.**

7) In the input box following **x**, type in "A2" (where the first random variable is located).

8) In the input box following **Mean**, type in 0.929.

9) In the **Cumulative** box, type in **"False"**.

10) Click on **OK**. You will now see the probability that a randomly selected region with an area of .25 square kilometers was hit exactly zero times. To match the results in the book, you can format your column to show 3 decimal places.

11) Using the fill handle, fill in the remaining cells.

12) Now suppose we want to use this probability to compute the Expected Number of Regions as found in Table 4-4 on page 212 in the book. In cell C1, type in "Expected Number of Regions". To have the text "wrap" to fit in to the formatted cell width, click on **Format, Cell,** then click on the **Alignment** tab, and click on **Wrap Text**.

13) Move to cell C2. We want to multiply each probability by the total number of regions (576). Enter the formula: = 576 * B2 into cell C2. Then use the fill handle to copy this formula down into the remaining cells. You will notice that some of the numbers generated are slightly different from those in the table in the book. This is because even when Excel is displaying only 3 decimal digits based on our cell formatting, it is using the longer string in computations.

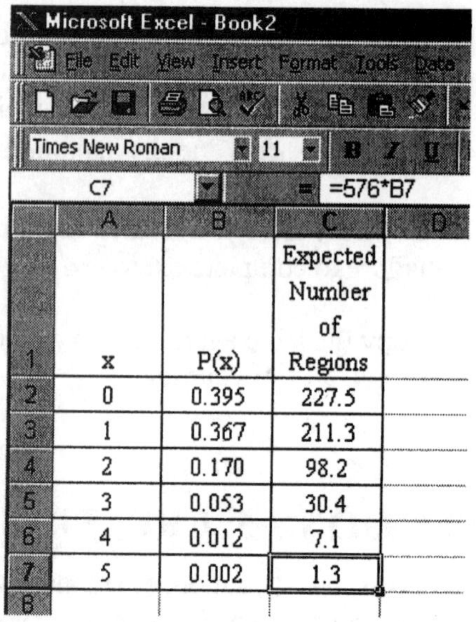

Microsoft Excel - Book2

File Edit View Insert Format Tools Data

Times New Roman 11

C7 =576*B7

	A	B	C
1	x	P(x)	Expected Number of Regions
2	0	0.395	227.5
3	1	0.367	211.3
4	2	0.170	98.2
5	3	0.053	30.4
6	4	0.012	7.1
7	5	0.002	1.3

TO PRACTICE THESE SKILLS

You can apply these technology skills by working on the following exercises. Make sure you save your work using a file name that is indicative of the material contained in your worksheets.

1) Create a table showing "x" and "P(x)" for exercise 5 on page 214 of your text book.

2) Create a table showing "x", "P(x)" and the "Expected Number" columns for exercise 9 on page 214 of your text book. In this exercise, you first need to compute the mean by dividing 116 by 365.

3) Create both the binomial distribution table and the Poisson distribution table for the conditions given in exercise 12 on page 215 of your book. Use your values to answer the questions posed.

CHAPTER 5: NORMAL PROBABILITY DISTRIBUTIONS

SECTION 5-1: OVERVIEW

In this chapter we will explore how to compute probabilities for a normal distribution, as well as find specific values if we are given information about the probability for a particular normal distribution. There are essentially five different functions that we can use when exploring normal distributions.

NORMSDIST: This function returns the **standard** normal cumulative distribution for a specified z value.

NORMSINV: This function returns the inverse of the standard normal cumulative distribution for a specified z value.

STANDARDIZE: This function returns a standardized score value for specified values of the random variable, mean and standard deviation.

NORMDIST: This function returns the cumulative normal distribution for specified values of the random variable, mean and standard deviation.

NORMINV: This function returns the inverse of the normal cumulative distribution for specified values for the probability, mean and standard deviation.

SECTION 5-2: WORKING WITH THE STANDARD NORMAL DISTRIBUTION

The first normal distribution presented in your text is the standard normal distribution. This distribution has a mean of 0 and a standard deviation of 1.

Finding P(0 < z < a)

Suppose we want to find the probability that a randomly selected z score is between 0 and 1.58. First we must recognize that Excel computes probabilities by determining the total area under the normal distribution **from the left up to a vertical line at a specific value**. Understanding this, we can see that we would first need to compute the probability that our value was less than 1.58, and from that value subtract the probability that our value was less than 0. This would leave us with the area under the curve between 0 and 1.58.

1) In cell A1, type in "x", and in cell B1, type in "P(z<x)" to indicate the probability that a z score is less than the particular x value.

2) In cell A2, type in 1.58.

3) Position your cursor in cell B2, and click on the **Function** icon in the toolbar or click on **Insert**, and click on **Function**.

4) In the **Function Category** box, click on **Statistical**, and in the **Function Name** box, click on **NORMSDIST**. Make sure you select the name with the "S" for Standard Normal. Then click on **OK** at the bottom of the dialog box.

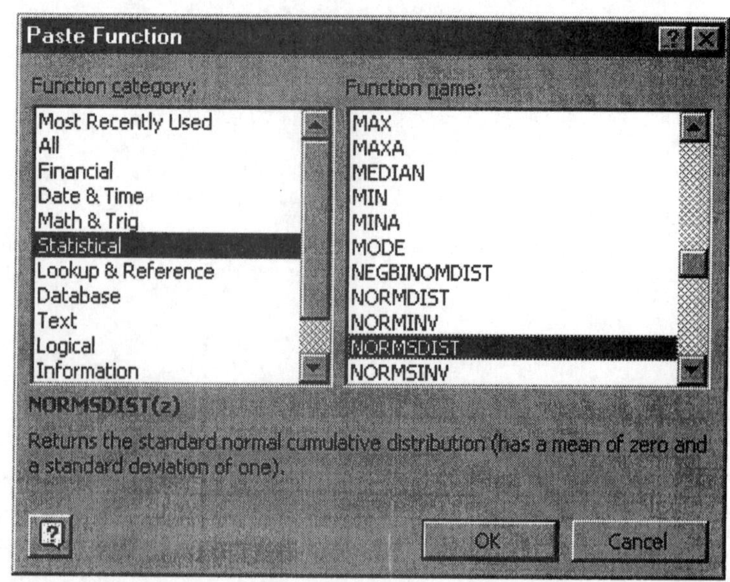

5) You will now see the **NORMSDIST** (Standard Normal Distribution) dialog box. In the entry box by the **z**, type in "A2" to indicate the cell where your z value is stored.

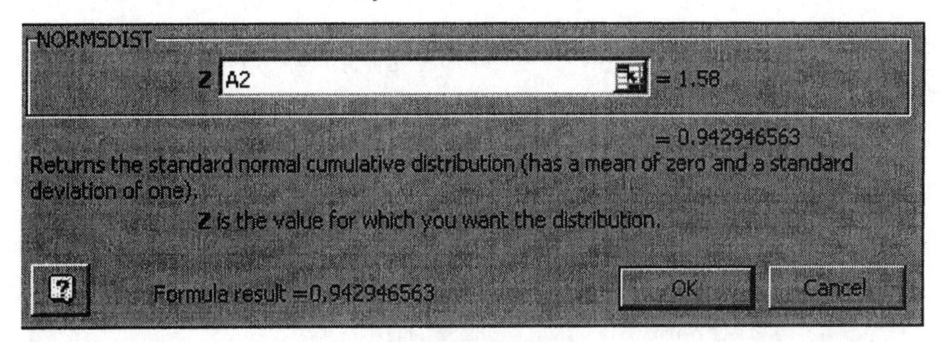

6) Click on **OK**. You should now see the value .942947 displayed in cell B2. This represents the probability that your randomly selected value is less than 1.58. (If you had formatted your column differently, you might have more or less decimal places displayed.)

7) In cell C1, type in "P(0<z<x)" to represent the probability that your score is between 0 and x.

8) In cell C2, enter the formula =b2-.5. The resulting value is the probability that you will randomly select a value that is between 0 and 1.58. (Since a normal distribution is symmetric about the mean, there is an area of .5 to the left of the mean of 0.)

9) Since we now have the formulas entered, we can easily expand our table to compute other probabilities that randomly selected values will fall between 0 and some value greater than 0. In column A, type in 1.2, and 2.3. Move to cell B2 and C2, and use the fill handle to fill in the remainder of the table.

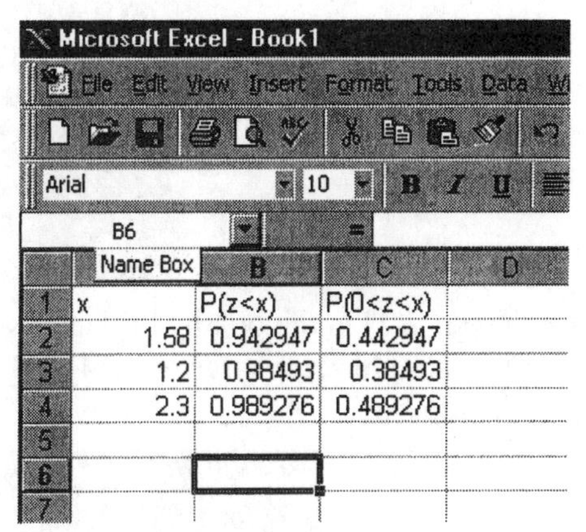

Finding P(a < z < b)

Suppose we want to find the probability that a randomly selected score is between 1.2 and 2.3. We can use the **NORMSDIST** function to find the $P(z < 1.2)$ and $P(z < 2.3)$. We can then set up a formula that would subtract the values $P(z < 2.3) - P(z < 1.2)$ to find the desired probability. Using the values from the above table, we should end up with a value of .104346

Finding P(a < z < 0) When a is Negative

Suppose we want to find the probability that a randomly selected score is greater than -1.3 but less than 0. We can use the **NORMSDIST** function to find $P(z < -1.3)$. Since we know that $P(z<0) = .5$ by the symmetry of the distribution, we can then subtract the value we produce for $P(z<-1.3)$ from .5 to find the desired probability.

Finding a Score When Given the Probability

Let's assume that we are working with thermometers that are normally distributed with a mean of 0 degrees Celsius and a standard deviation of 1 degree Celsius. Suppose we want to find the 95th percentile. This means that we want the area to the left of our value to be .95.

1) Type "P(z < x)" in cell A1, "x" in cell B1, and .95 in cell A2.

2) Position your cursor in cell B2, and click on the **Function** icon on the tool bar (or click on **Insert, Function**).

3) Click on **Statistical, NORMSINV** in the **Function Category** and **Function Name** boxes respectively. Then click on **OK**.

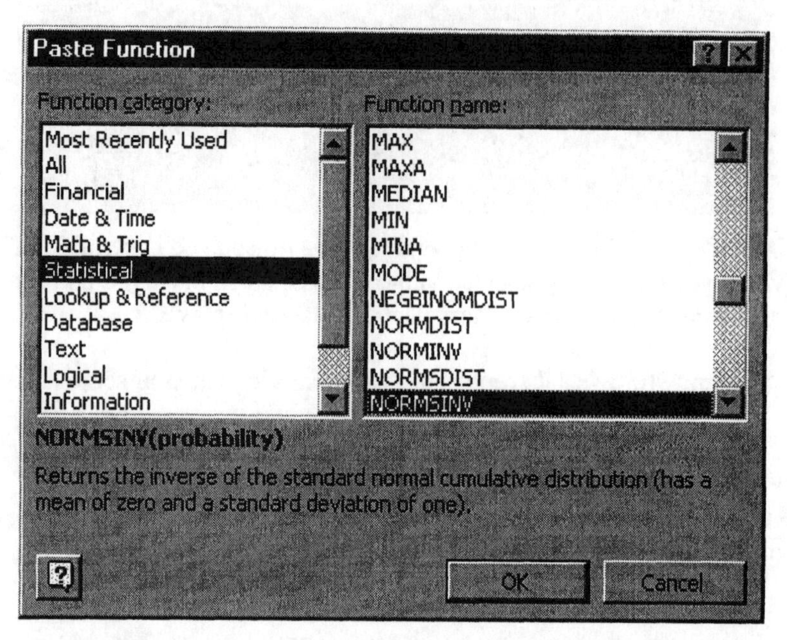

4) Type "A2" in the box by Probability to indicate that the probability is found in this cell, and click on **OK**. You should see a value of 1.644653, which is the score separating the bottom 95 % of the scores from the top 5 %.

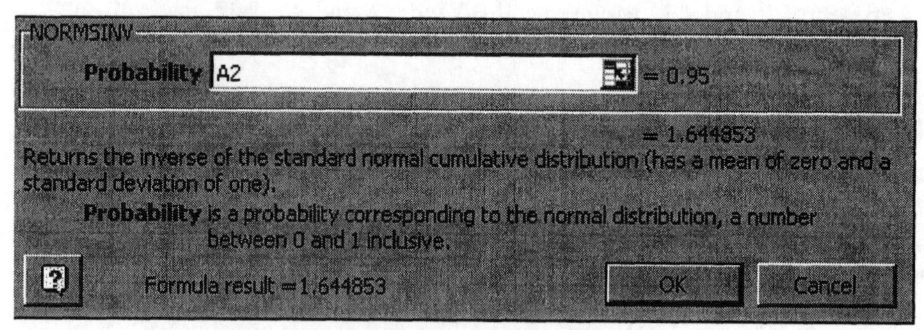

5) You can now enter other areas to the left of the value you want to find, and then find the corresponding z scores by filling the column B with the formula from cell B2.

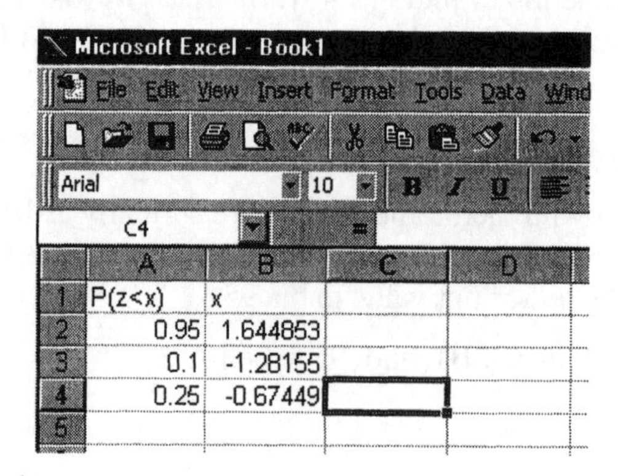

TO PRACTICE THESE SKILLS

You can apply the skills you learned in this section by using Excel to complete the odd numbered exercises from 9 - 39 starting on page 239 in your textbook.

SECTION 5-3: WORKING WITH NON-STANDARD NORMAL DISTRIBUTIONS

Standardizing Scores

If we are working with a normal distribution which is not a standard normal, we could opt to "standardize" the scores, and then use the techniques for finding probabilities and values as presented in the Standard Normal section.

Let's work with the example for **Jet Ejection Seats** on page 242 of your text. This example states that women's weights are normally distributed with a mean of 143 lb and a standard deviation of 29 lb. Suppose we want the probability that if a woman is randomly selected, she weighs between 143 and 201 lb.

1) In cell A1, type in "x"; in B1, type in "z"; in C1, type in "Mean"; in D1, type in "SD".

2) In cell A2, type in 201; in C2, type in 143; and in D2, type in 29.

3) Position your cursor in cell B2. Click on the **Function** icon on the tool bar, and click on **Statistical**, **STANDARDIZE.** Then click on **OK**.

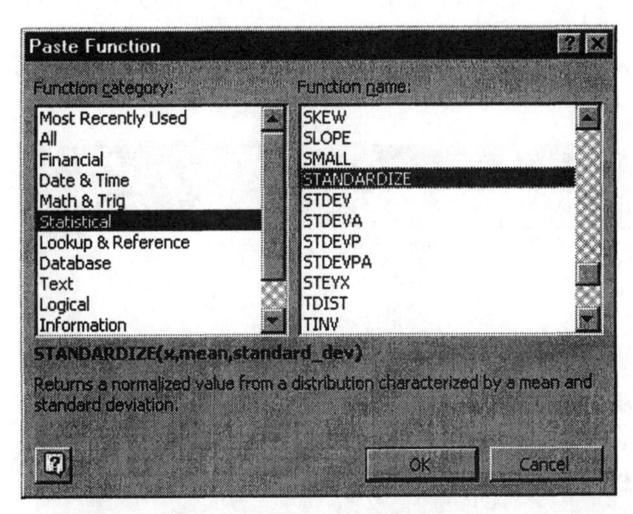

4) In the **Standardize** dialog box, type in "A2" to indicate where the **x** value is stored. Type in "\$C\$2" to indicate that the **Mean** is an absolute address, and type in "\$D\$2" to indicate that the **Standard _dev** is also an absolute address. Then click on **OK**.

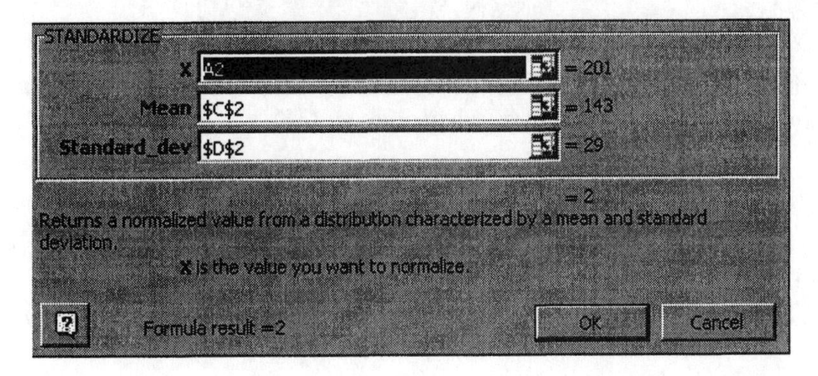

5) The value produced is the "Standardized score" and represents how many standard deviations the value 201 is away from the mean of 143. You can now use the **NORMSDIST** function to compute probabilities for these z scores, as covered in the previous section.

6) To standardize other scores, type the scores in column A, and use the fill handle to copy the formula down in column B.

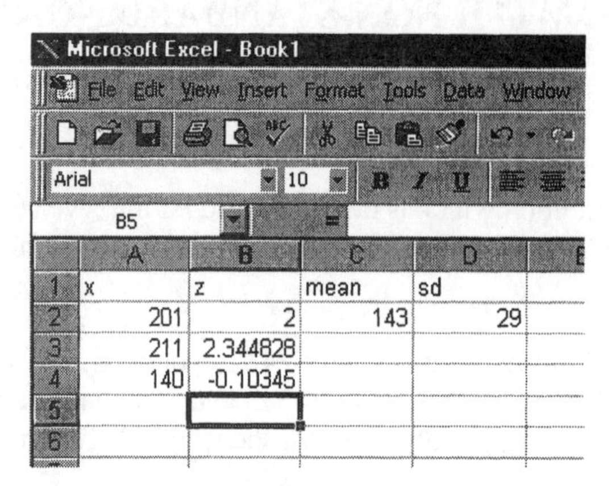

Finding Probabilities Using the NORMDIST Function

We don't need to standardize the scores to find probabilities for non-standard normal distributions.

1) Type in the values 201, 211 and 140 in column A, titling the column "a".

2) Type in "P(x < a)" in cell B1, and position your cursor in cell B2.

3) Click on the **Function** icon from the tool bar, and click on **Statistical, NORMDIST** from the **Function Category** and **Function Name** respectively. Then click on **OK**.

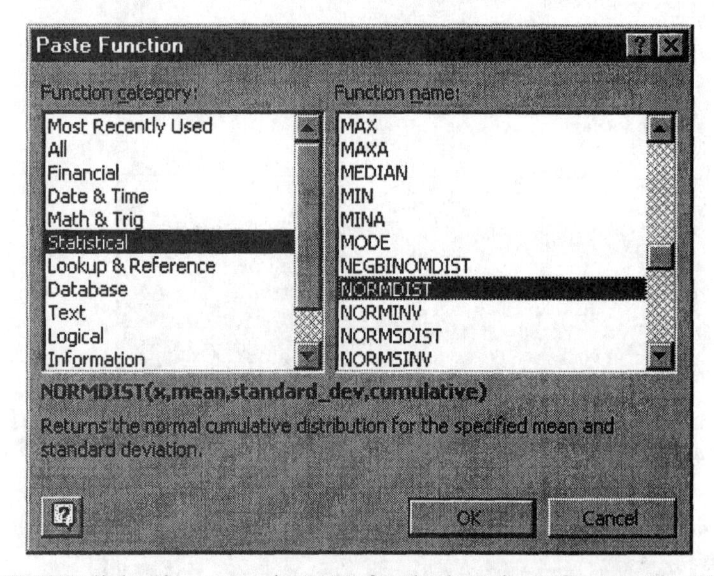

4) In the **NORMDIST** dialog box, type in "A2" for the location of **x**, type in 143 for the **Mean**, 29 for the **Standard_dev**, and type "True" for Cumulative. Then click on **OK**.

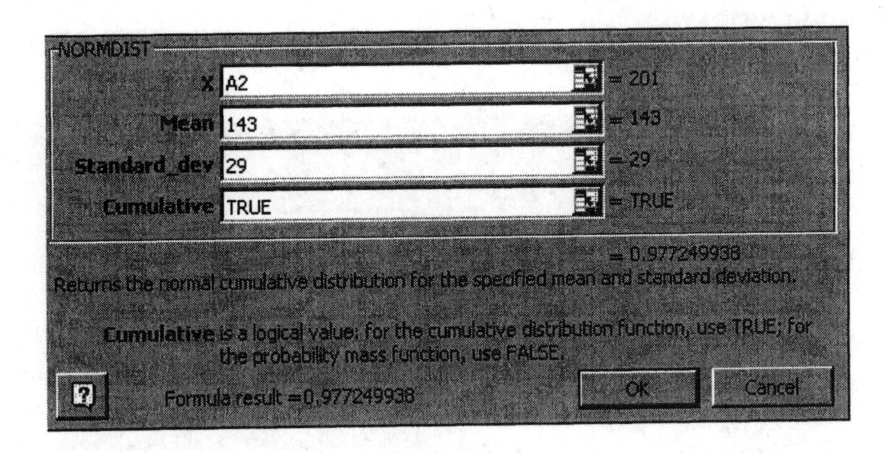

5) Remember that the resulting value represents the probability that a randomly selected woman weighs less than 201 lb. If you now wanted to find the probability that a randomly selected woman weighed between 143 and 201 lbs, you would need to subtract the area to the left of the mean (.5) from the generated probability. This would produce: $P(143 < x < 201) = .47725$.

6) You can now create other probabilities $P(x < a)$ by typing in appropriate values in column A and then using the fill handle to create the probabilities in column B. If the value a is above the mean, you can create the value $P(\text{mean} < x < a)$ by subtracting .5 from the resulting value.

7) Suppose we want $P(140 < x < 201)$. We can see in the table above that $P(x < 211) = .990482$ and that $P(x < 140) = .458804$. Therefore, $P(140 < x < 211) = P(x < 211) - P(x < 140) = .990482 - .458804 = .531679$.

TO PRACTICE THESE SKILLS

You can use the functions outlined in this section to complete the odd numbered exercises from 5 through 23 starting on page 248 of your textbook.

SECTION 5-4: FINDING VALUES FOR NORMAL DISTRIBUTIONS

Suppose we want to find the 10th percentile for women's weights. Suppose we know the weights are normally distributed with a mean of 143 lb and a standard deviation of 29 lb.

Remember, Excel works with the area to the left of the value that we want.

1) In cell A1, type "P(X < x)". In cell B1 type "x". In cell C1, type "Mean". In cell D1, type "Std. Dev". In cell A2, type in .10 to represent that there is an area of .10 to the left of the desired score. In cell C2, type in 143. In cell D2, type in 29.

2) Position your cursor in cell B2, and click on the **Function** icon on the tool bar (or click on **Insert, Function**).

3) Click on **Statistical, NORMINV** in the **Function category** and **Function name** boxes respectively. Then click on **OK**.

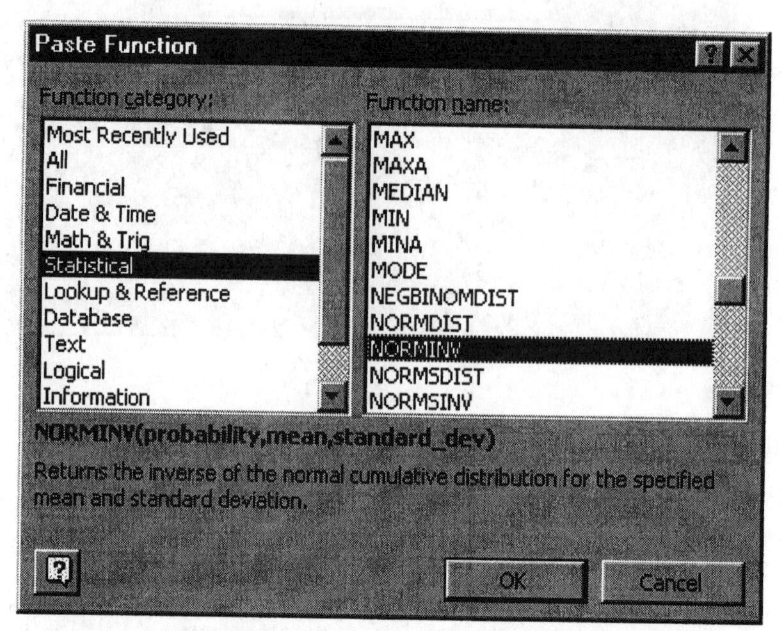

4) In the **NORMINV** dialog box, type in "A2" in the box after **Probability** to indicate this is where the probability is stored. In the **Mean** box, type in the absolute address "$C$2", and in the **Standard_dev** box, type in the absolute address "$D$2". (Remember, an absolute address entry will not be automatically updated when you fill a column.) Then click on **OK**.

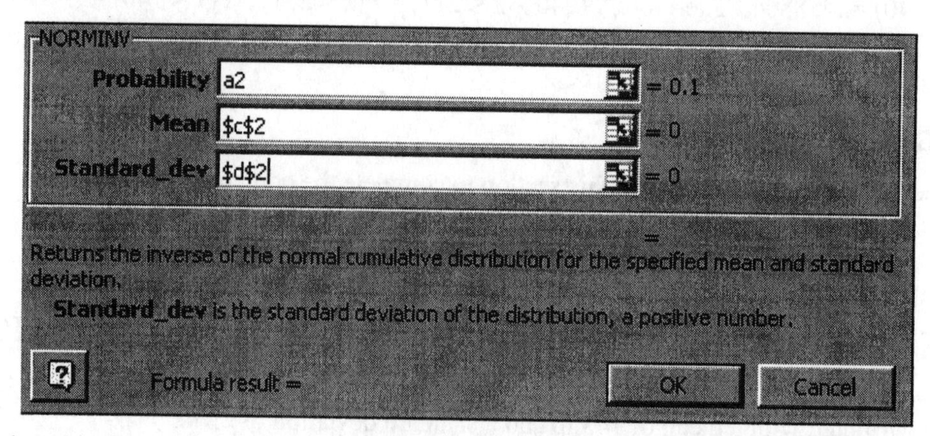

5) The value produced is the score which separates the bottom 10% of the data from the top 90%.

6) In column A, you can now type in other areas to the left of values for percentiles you are interested in, and use the fill handle to fill column B with the formula from B2. Try finding the remaining Deciles.

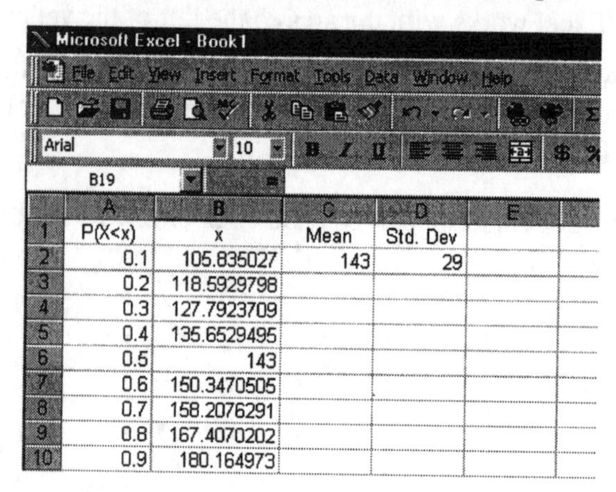

	A	B	C	D	E
1	P(X<x)	x	Mean	Std. Dev	
2	0.1	105.835027	143	29	
3	0.2	118.5929798			
4	0.3	127.7923709			
5	0.4	135.6529495			
6	0.5	143			
7	0.6	150.3470505			
8	0.7	158.2076291			
9	0.8	167.4070202			
10	0.9	180.164973			

TO PRACTICE THESE SKILLS

Once you have the table above set up, you can quickly change the mean, standard deviation and probability and Excel will update the value in the "x" column.

Try using the table you set up, and modify the numbers as appropriate to address the odd exercises 1 - 19 starting on page 252 of your textbook.

After you complete an exercise, you may want to copy and paste your completed table to another location in your worksheet. After you have copied the table, you should activate the cell where you want your table to begin. Then click on **Edit, Paste Special**, and click in the bubble by the word **Values**. Then click on **OK**. If you activate one of the cells where you had entered a formula originally, you will notice that now only the value shows up. The original formula is no longer active in the copied table. You can go back to the original table to compute the value for the next problem.

SECTION 5-5: THE CENTRAL LIMIT THEOREM

In this section, we will use Excel to help us visualize the Central Limit Theorem. We will create a model similar to the one in your textbook, but by using the power of Excel, we can easily consider a higher sample size. We will create a table of 10 columns of randomly generated digits, each column containing 50 values.

Refer back to the instructions in section 3-3 of this manual on "Generating Random Numbers".

1) In your worksheet, type "SSN Digits" in cell A1.

2) Position your cursor in cell A2, and follow the instructions to create a random number between 0 and 9 in that cell.

3) Use the fill handle to fill this formula to cells B2 through J2.

4) Use the fill handle again to fill a table down to row 51. You should now have a table with 500 digits in it. Your table should look different from others, in that the values in each cell are being randomly generated. Just be aware of this if you are comparing your table to another classmate's table.

5) We want to find the mean of each row. Position your cursor in cell L1 and type in "ROW MEAN".

6) Position your cursor in cell L2, and click on the function icon in the toolbar. In the **Function category** box, click on **Statistical**, and in the **Function name** box, click on **Average.** Click on **OK**.

7) In the input box by Number 1, type in the range "A2:J2", and click on **OK**. In cell L2, you should now see the average of the ten numbers in row 2.

8) You may notice that as you work with your columns and cells, the table of numbers you generated changes. To stabilize the set of data you will be working with, highlight the columns containing your values and your means and click on **Edit, Copy.**

9) Move to a new worksheet and position your cursor in cell A1. From the toolbar, click on **Edit, Paste Special**, and then click on the bubble by **Values**. You will now have a stable table, since the cells are representing numbers now rather than formulas. Your pasted table may very well be quite different from the table that you chose to copy!

10) Create a Histogram for the 500 digits found in your table. (Refer back to Chapter 2 if necessary.)

11) You should see clearly that your histogram does not appear bell shaped.

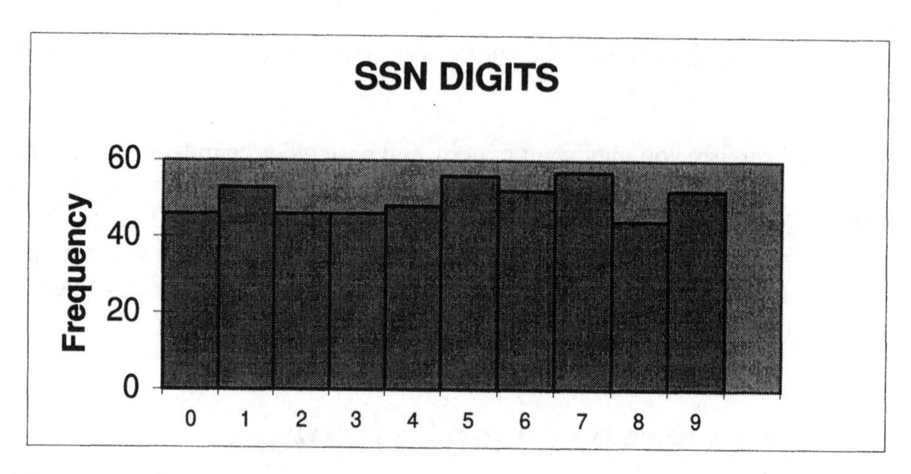

12) Now create a histogram of your 50 sample means.

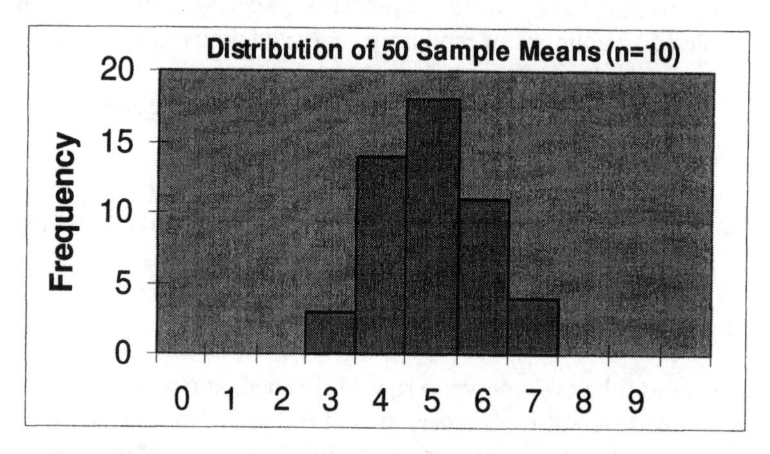

13) Although this may not look entirely bell shaped, it is definitely moving in that direction. Keep in mind that our sample size is still quite small. If we increased the number of values we used to compute each sample mean, we should find that our distribution of sample means becomes increasingly bell shaped.

Computing Probabilities For Situations Involving the Central Limit Theorem

In section 5.3, we learned how to use the **NORMDIST** function to compute probabilities for non-standard normal distributions. In a situation involving the Central Limit Theorem, we will use the same approach, however we will need to enter the standard deviation as $\sigma/\sqrt{n}$.

TO PRACTICE THESE SKILLS

You can apply the skills you learned in this section by working on the following exercises.

1) Repeat the exercise presented in this section, but using a table with 100 rows of 10 digits. Compare the histograms you create from your new data set to those presented in this section.

2) Many of the exercises in section 5-5 of your textbook deal with finding probabilities. Use the patterns established in the previous section to complete the odd numbered exercises 1 - 19, starting on page 263 of your textbook. Notice that when you are looking for the probability involving a mean, you must recognize that the Central Limit applies, and that the standard deviation you must use in generating your probability will be $\sigma/\sqrt{n}$.

SECTION 5-6: NORMAL DISTRIBUTION AS APPROXIMATION TO BINOMIAL

If we have a binomial distribution where $np \geq 5$ and $nq \geq 5$, we can approximate binomial probability problems by using a normal distribution. The material below helps you see a clear demonstration that this approach will work. We will plot three different binomial probability distributions to see that as the sample size increases, our distribution appears to look more and more like a normal distribution.

In a new worksheet, we will create three binomial probability distributions:

- Let the first distribution have n = 10 and p = .5
- Let the second distribution have n = 25 and p = .5
- Let the third distribution have n = 50 and p = .5

1) Refer back to section 4-3 on Binomial Distributions and create your pairs of columns for the random variable and the associated probabilities in columns A& B; D & E; and G & H.

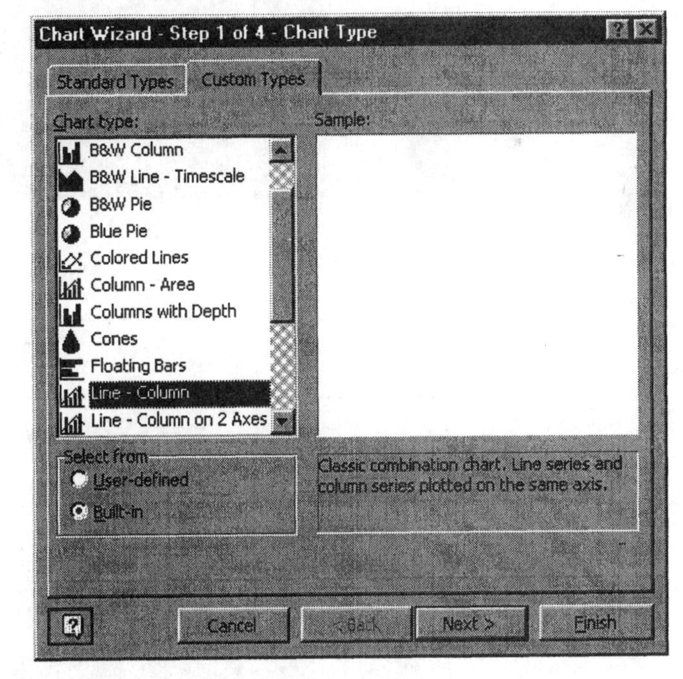

2) Click on the **Chart** icon, and click on the **Custom Types** tab. Scroll down until you see the **Line-Column** option. Click on this option.

3) Click on **Next**.

4) For your first graph, select the cells containing the probabilities in column B.

5) Click on the **Series** tab, and underneath the **Series** box, click on **Add**. You should see another Series name in the box.

6) Delete the information that is in the **Values** box, and select the probabilities that are in column B.

7) Click on **Next,** and enter the appropriate information for the Chart Title and the axes names.

8) Click on **Next.** Choose to insert the chart in a new sheet.

9) Once your chart appears, double click in the **Plot Area**. Click on **Options** in the dialog box, and change your gap size to 0.

10) Repeat this process to create the pictures for the other two binomial distributions. Your pictures should appear as those below:

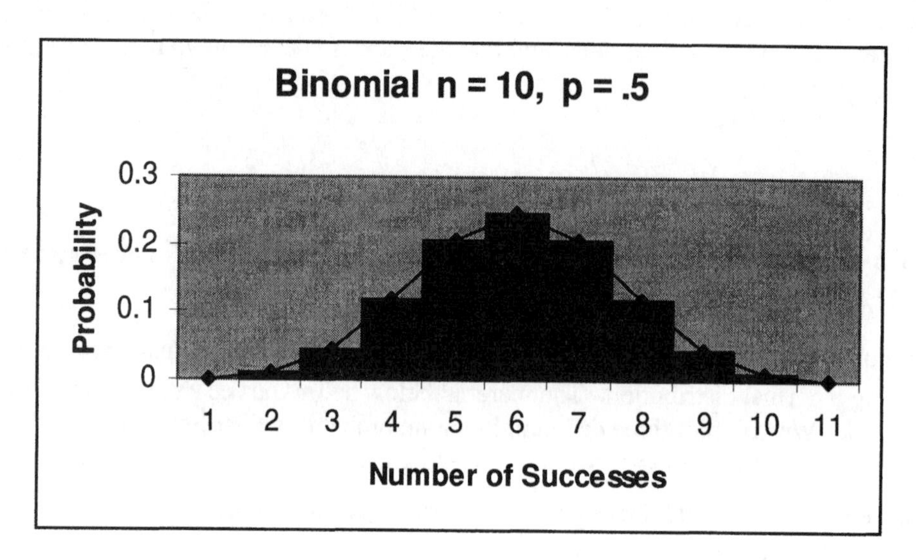

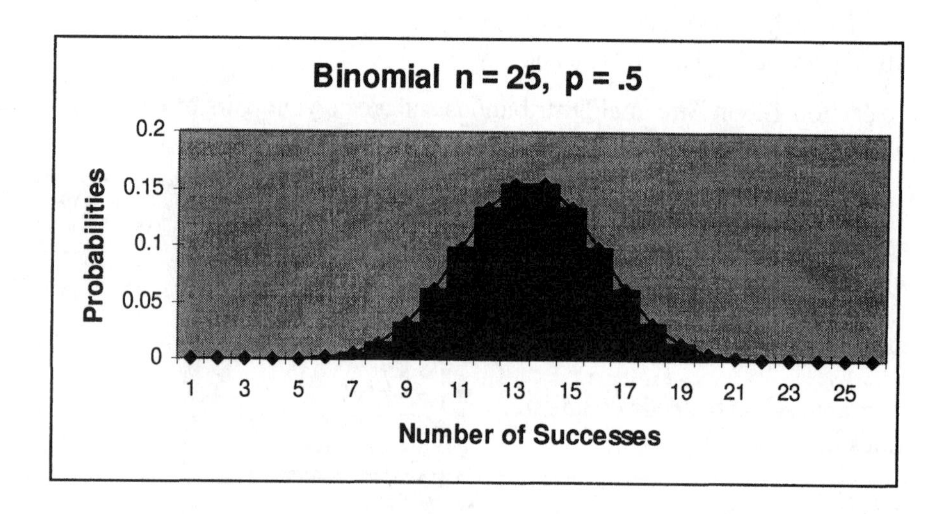

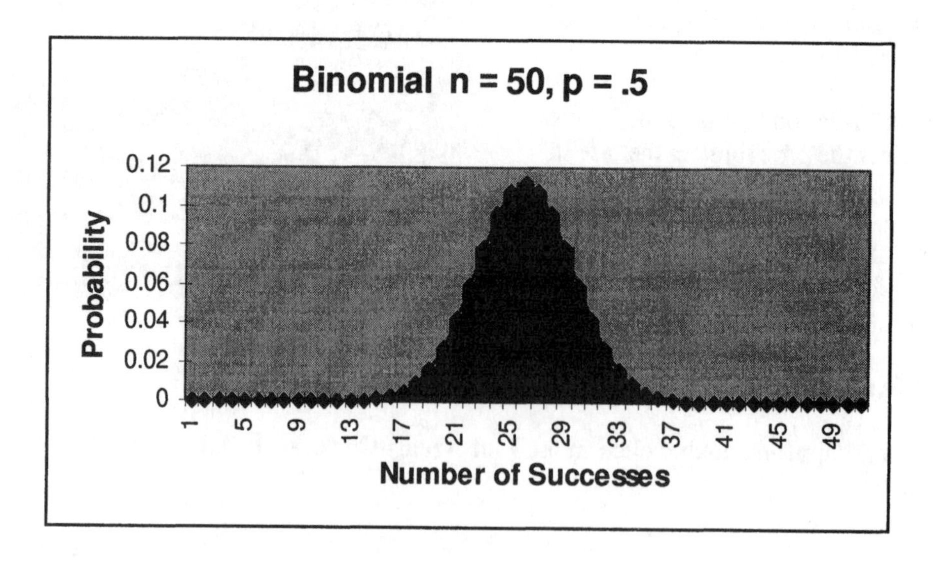

11) What you should notice about each of these successive pictures is that the areas in the bars, which represent the probabilities for each random variable, more closely approximates the area contained under the curve. If you think of the curve as representing a normal curve, than you can clearly see that the binomial probability distribution is more closely approaching a normal distribution.

We can see that the probabilities we find using the normal distribution will closely match those that we can generate from the binomial distribution. To see this clearly, we will use the case of $n = 50$ and $p = 0.5$.

1) Using the values in our binomial distribution table, we can find $P(x \geq 30)$ by adding up the probabilities from $x = 30$ to $x = 50$. This produces a value of .101319.

2) If we want to use a normal distribution, we will need to compute the mean and standard deviation using the formulas for a binomial distribution. You should find the value for the mean is 25 and the standard deviation is 3.535534.

3) As outlined in the text on page 272, we need to use a continuity correction when using the normal distribution to approximate the binomial. Since we want to find the probability of getting a value greater than or equal to 30, we should find $P(x \geq 29.5)$. Use the **NORMDIST** function on Excel. Remember that Excel returns a value that represents the area under the normal distribution curve to the left of the value. In order to find the probability we really want, we will need to subtract this value from 1. The **NORMDIST** function returns a value of 0.898454. Subtracting this value from 1 produces a probability of .101546. Recall that the probability using our binomial information is .101319. If we were to increase our sample size, we would find that we get even closer approximations when using the normal distribution to approximate the binomial.

TO PRACTICE THESE SKILLS

You can apply the skills you learned in this section by working on the following exercises.

1) Create the picture showing the Line-Column graph for a binomial probability distribution where $n = 100$ and $p = 0.5$. Compare your picture to those for $n = 10$, 25 and 50 found in this section.

2) Use the **NORMDIST** function on Excel, and the continuity correction to find the probabilities for the odd numbered exercises 1 - 7 starting on page 275 of your text book. You may want to refer back to instructions found in section 5-3 of this manual.

3) Use the **BINOMDIST** function on Excel, and, when appropriate, the **NORMDIST** function with the continuity correction to find the probabilities asked for in the odd numbered exercises 9 - 27 starting on page 276 of your book. You may want to refer back to instructions found in section 4-3 of this manual.

SECTION 5-7: DETERMINING NORMALITY

Oftentimes we want to know whether the data we are working with is normally distributed. We have already learned how to create histograms for sample data. From our histogram, we can reject normality if the histogram departs dramatically from bell shape.

An alternate way to determine normality is to construct a normal probability plot for the data. In a normal probability plot, the observations in the data set need to be ordered from smallest to largest. These values are then plotted against the expected z scores of the observations calculated under the assumption that the data are from a normal distribution. When the data are normally distributed, a linear trend will result. A nonlinear trend suggests that the data are non-normal.

We can use the **DDXL** Add-In to generate a normal probability plot. We will create a model similar to the **Diet Pepsi** example found in your book on page 281, but using the sample of 36 weights of diet Coke listed in Data Set 1 of Appendix B.

1) Load the data from the CD that comes with your book (COLA.XLS) Copy the column showing the 36 weights of Diet Coke into column A of a new Excel worksheet.

2) Click on **DDXL** on your menu bar.

3) Click on **Charts and Plots**. Click on the down arrow on the Function type box, and click on **Normal Probability Plot.**

4) Click on the pencil icon for **Quantitative Variable**, and enter the range of values for your data. If you started in cell A1, you range will be entered as "A1:A36".

5) Click on **OK.** You should see the following information on your screen.

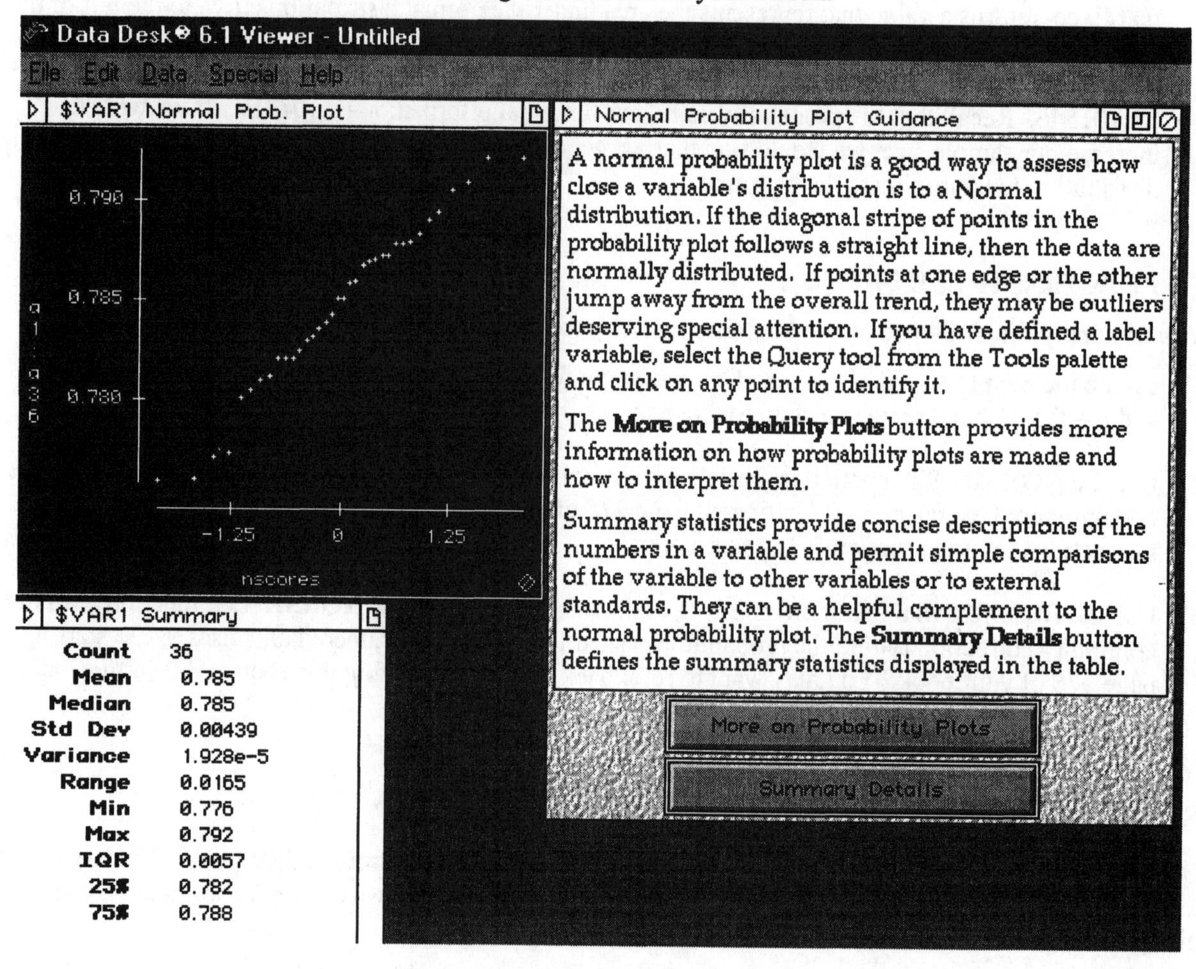

TO PRACTICE THESE SKILLS

You can apply the skills from this section by working with exercises 5, 6, 7 & 8 on page 283 of your text book. For each exercise, you should open the file in an Excel workbook from the CD that comes with your book. The file names are listed below:

For exercise 5, Boston Rainfall data is in the file named BOSTRAIN.XLS.
For exercise 6, Weights of M&Ms data is in the file named M&M.XLS.
For exercise 7, Old Faithful Geyser data is in the file named OLDFAITH.XLS.
For exercise 8, Weights of sugar Packets data is in the file named SUGAR.XLS

CHAPTER 6: CONFIDENCE INTERVALS

SECTION 6-1: OVERVIEW

Many times we do not know the value of the parameters that are used to describe a population and need to resort to using information contained in a sample. If we can identify a numerical value that describes the sample, then this value can also be used to estimate the corresponding descriptor for the population. **Confidence intervals** are important in statistics because they allow you to gauge how accurately a sample parameter approximates the same parameter with respect to the population. The confidence interval gives you a range of values and a probability. The probability value tells you the likelihood that you have an interval that actually contains the value of the unknown population parameter. Components of a confidence interval include a lower limit and an upper limit for the parameter under consideration and as well as a probability value.

Excel does not have a built in function that automatically calculates the confidence interval. We will need to rely on some of the functions we have already used in Excel to help us with this process. In addition to exploring this topic with Excel we will make use of the add-in DDXL which you added to your menu bar in chapter 2.

CONFIDENCE: Returns the margin of error or maximum error for a population mean. **CONFIDENCE(alpha, standard_dev,size)** where alpha refers to the significance level used to compute the confidence interval, standard_dev is the population standard deviation for the data range and size is the sample size.

TINV: Returns the inverse of the Student's t-distribution for the specified degrees of freedom. **TINV(probability, deg_freedom)** where probability is the probability associated with a two tailed Student's t distribution and deg_freedom is a positive integer indicating the number of degrees of freedom needed to characterize the distribution.

SECTION 6-2: ESTIMATING A POPULATION MEAN: LARGE SAMPLES

The formula for the confidence interval for the population mean when the standard deviation σ is known is given by $\overline{X} - E < \mu < \overline{X} + E$ where $E = z_{\alpha/2} \cdot \frac{\sigma}{\sqrt{n}}$

To determine a confidence interval for the mean in Excel you must know the value for $\overline{X}$ and for σ. These can be found using the built in statistical functions within Excel or from Descriptive Statistics in the Data Analysis Tools.

Begin by entering the data found in the Chapter Problem at the beginning of Chapter 6 (Table 6 –1, Body Temperatures of 106 Healthy Adults) into Excel.

Use the **Function** icon to determine the sample mean and standard deviation.

DETERMINING CONFIDENCE INTERVALS

To find the **confidence interval** we need to use the **Paste Function** feature.

1) Under Function category click on **Statistical.**

2) Scroll through the list of Function names until you see **CONFIDENCE.** Click on this function name.

3) Click **OK.**

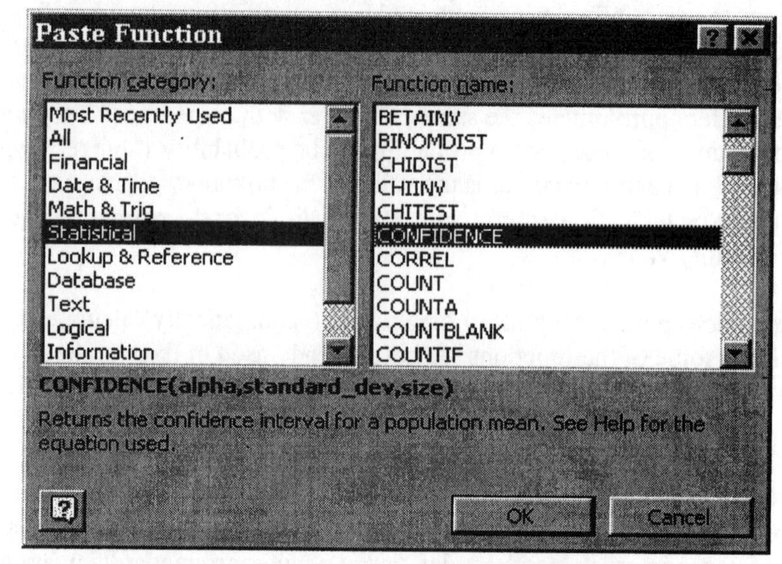

The following dialogue box will appear in the upper left-hand corner of your Excel workbook.

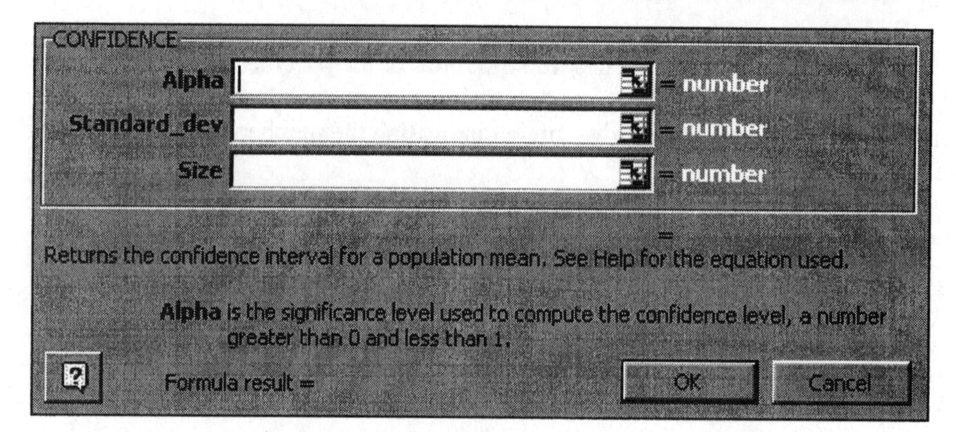

4) Assuming a degree of confidence of 95% we can fill in the information required. Recall that with a degree of confidence of 95%, $\alpha = 0.05$. Enter the standard deviation and the size of the sample. Your worksheet should look similar to the following screen.

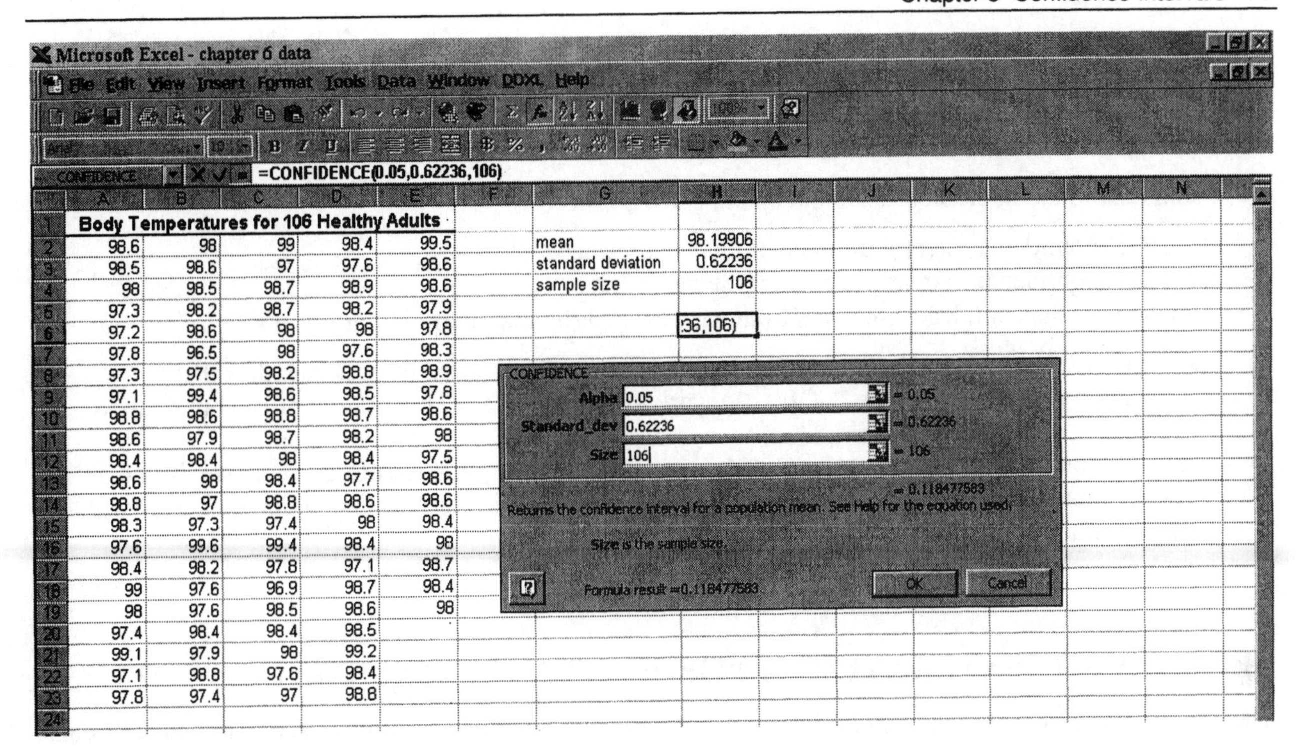

	A	B	C	D	E	F	G	H
1	Body Temperatures for 106 Healthy Adults							
2	98.6	98	99	98.4	99.5		mean	98.19906
3	98.5	98.6	97	97.6	98.6		standard deviation	0.62236
4	98	98.5	98.7	98.9	98.6		sample size	106
5	97.3	98.2	98.7	98.2	97.9			
6	97.2	98.6	98	98	97.8			36,106)
7	97.8	96.5	98	97.6	98.3			
8	97.3	97.5	98.2	98.8	98.9			
9	97.1	99.4	98.6	98.5	97.8			
10	98.8	98.6	98.8	98.7	98.6			
11	98.6	97.9	98.7	98.2	98			
12	98.4	98.4	98	98.4	97.5			
13	98.6	98	98.4	97.7	98.6			
14	98.8	97	98.8	98.6	98.6			
15	98.3	97.3	97.4	98	98.4			
16	97.6	99.6	99.4	98.4	98			
17	98.4	98.2	97.8	97.1	98.7			
18	99	97.6	96.9	98.7	98.4			
19	98	97.6	98.5	98.6	98			
20	97.4	98.4	98.4	98.5				
21	99.1	97.9	98	99.2				
22	97.1	98.8	97.6	98.4				
23	97.8	97.4	97	98.8				
24								

The value returned is called the **margin of error** (or **maximum error**) and is denoted by the letter E in the formula presented at the beginning of this section.

1) Determine the upper confidence interval limit ($\overline{X}+E$) and the lower confidence interval limit ($\overline{X}-E$).

2) Using the general format $\overline{X}-E<\mu<\overline{X}+E$ used to display a confidence interval substitute in the values found in step #1 above.

   a) The lower limit is found using the formula =(H2-H6)

   b) The upper limit is found by using the formula =(H2+H6)

3) This will give you a confidence interval $98.08^{\circ}<\mu<98.32^{\circ}$.

Microsoft Excel - chapter 6 data

File Edit View Insert Format Tools Data Window DDXL Help

H10

	A	B	C	D	E	F	G	H
1	Body Temperatures for 106 Healthy Adults							
2	98.6	98	99	98.4	99.5		mean	98.19906
3	98.5	98.6	97	97.6	98.6		standard deviation	0.62236
4	98	98.5	98.7	98.9	98.6		sample size	106
5	97.3	98.2	98.7	98.2	97.9			
6	97.2	98.6	98	98	97.8		margin of error	0.118478
7	97.8	96.5	98	97.6	98.3			
8	97.3	97.5	98.2	98.8	98.9		lower limt	98.08058
9	97.1	99.4	98.6	98.5	97.8		upper limit	98.31753
10	98.8	98.6	98.8	98.7	98.6			
11	98.6	97.9	98.7	98.2	98			
12	98.4	98.4	98	98.4	97.5			
13	98.6	98	98.4	97.7	98.6			
14	98.8	97	98.8	98.6	98.6			
15	98.3	97.3	97.4	98	98.4			
16	97.6	99.6	99.4	98.4	98			
17	98.4	98.2	97.8	97.1	98.7			
18	99	97.6	96.9	98.7	98.4			
19	98	97.6	98.5	98.6	98			
20	97.4	98.4	98.4	98.5				
21	99.1	97.9	98	99.2				
22	97.1	98.8	97.6	98.4				
23	97.8	97.4	97	98.8				
24								

TO PRACTICE THESE SKILLS

You can practice the technology skills learned in this section by working through the following problems in your textbook.

1) Using data found in problem 17 found on page 310 construct
   a) a 99% confidence interval estimate of the mean value
   b) a 90% confidence interval estimate of the mean value
   c) Compare the confidence intervals from parts (a) and (b) and interpret your results.
2) Use the data found in Data Set 1 in Appendix B of your textbook or on the CD data disk that accompanies your text in the file "COLA" to work through problem 21 on page 311 of your textbook.
3) Problem 25 on page 311 gives you an excellent opportunity to explore the effects outliers have on a confidence interval. Use the data already entered into Excel at the start of this section that deals with the body temperature of 106 healthy adults to work through this problem.

SECTION 6-3: ESTIMATING A POPULATION MEAN: SMALL SAMPLES

The student t distribution is used to estimate the population mean when the sample size is less than or equal to 30 ($n \leq 30$). The small sample confidence interval is $\overline{X} - E < \mu < \overline{X} + E$ where $E = t_{\alpha/2} \cdot \frac{s}{\sqrt{n}}$.

Note that the sample standard deviation replaces the population standard deviation in the formula. To determine the small sample confidence intervals or the population mean with Excel, use the **TINV** function to determine the appropriate ***t* values** .

Begin by entering the following data into an Excel worksheet.

Time needed to assemble an item (in minutes)				
5.2	5.7	4.6		
5.7	4.7	4		
5.2	4.3	4.9		
4.4	4.5	4.5		
4.8	4.6	5.8		
6	4.5			
5	4.8			
5.2	4			
5.6	4.8			

We will use this information to determine the average number of minutes it takes to manufacture one item. Use a 95% degree of confidence.

To find the confidence interval for μ, we need to
   a) Calculate the mean.
   b) Calculate the small sample standard deviation.

These can be determined easily using the built in functions in Excel.

DETERMINING CONFIDENCE INTERVAL USING A T DISTRIBUTION

To determine the ***t* value** we will follow the same method used to find the confidence interval outline earlier in this section:

To find the **TINV confidence interval** we need to use the **Paste Function** feature.

1) Under **Function** category click on **Statistical.**

2) Scroll through the list of **Function** names until you see **TINV**. Click on this function name.

3) Click **OK.**

The following dialogue box will appear in the upper left hand corner of your Excel workbook.

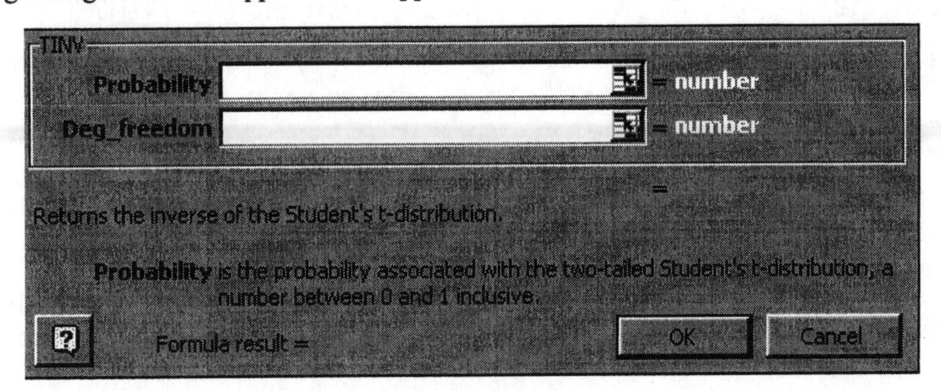

Assuming a degree of confidence of 95% we can fill in the information required. Recall that with a degree of confidence of 95%, the probability = 0.05, the deg_freedom is 24

Your worksheet should look similar to the screen below.

TINV =TINV(.05,24)

	A	B	C	D	E	F	G	H	I	J	K
1	Time needed to assemble an item (in minutes)										
2	5.2	5.7	4.6				mean	4.904348			
3	5.7	4.7	4				standard deviation	0.565266			
4	5.2	4.3	4.9								
5	4.4	4.5	4.5				t distribution	(.05,24)			
6	4.8	4.6	5.8				error				
7	6	4.5									
8	5	4.8					lower limit				
9	5.2	4					upper limit				
10	5.6	4.8									

TINV
Probability .05 = 0.05
Deg_freedom 24 = 24
= 2.063898137
Returns the inverse of the Student's t-distribution.
Deg_freedom is a positive integer indicating the number of degrees of freedom to characterize the distribution.
Formula result =2.063898137
OK Cancel

You can complete the information on the worksheet using the same procedures outlined for the confidence interval to get the following:

	A	B	C	D	E	F	G	H	I
1	Time needed to assemble an item (in minutes)								
2	5.2	5.7	4.6				mean	4.904348	
3	5.7	4.7	4				standard deviation	0.565266	
4	5.2	4.3	4.9						
5	4.4	4.5	4.5				t distribution	2.063898	
6	4.8	4.6	5.8				error	0.23333	
7	6	4.5							
8	5	4.8					lower limit	5.137678	
9	5.2	4					upper limit	4.671018	
10	5.6	4.8							
11									

TO PRACTICE THESE SKILLS

You can practice the technology skills learned in this section by working through the following problems in your textbook.

1) Use the data comparing yellow and brown M&M candy found in Data Set 10 in Appendix B of your textbook or on the CD data disk that accompanies your text in the file "M&M" to work through problem 21 on page 322 of your textbook.

2) Repeat the preceding problem using the Excel function CONFIDENCE (in other words treat this problem as if $n \geq 30$). How are the confidence interval limits affected if this function is incorrectly used instead?

SECTION 6-4: ESTIMATING A POPULATION PROPORTION

Excel does not produce confidence interval estimates for proportions. It will be necessary for us to use the **DDXL** add-in that came as a supplement to your textbook to do this. If you have not loaded **DDXL** please do so know. The instructions for adding this feature to Excel can be found in Chapter 2 of this manual.
If you are unsure as to whether or not you have already added DDXL check the menu bar in Excel to see if it featured on the menu bar as seen below.

File Edit View Insert Format Tools Data Window DDXL Help

DETERMINING A CONFIDENCE INTERVAL FOR A POPULATION PROPORTION:

To determine the confidence interval for a population proportion using **DDXL** we will begin by considering the example **Misleading Survey Responses** found on page 332 of your textbook.

1) Label cells A1 and A2 as shown in the Excel worksheet on the next page. Enter the number of successes in cell B1 and the number of trials in cell B2.

2) Click on **DDXL.**

4) Highlight **Confidence Intervals** and click.

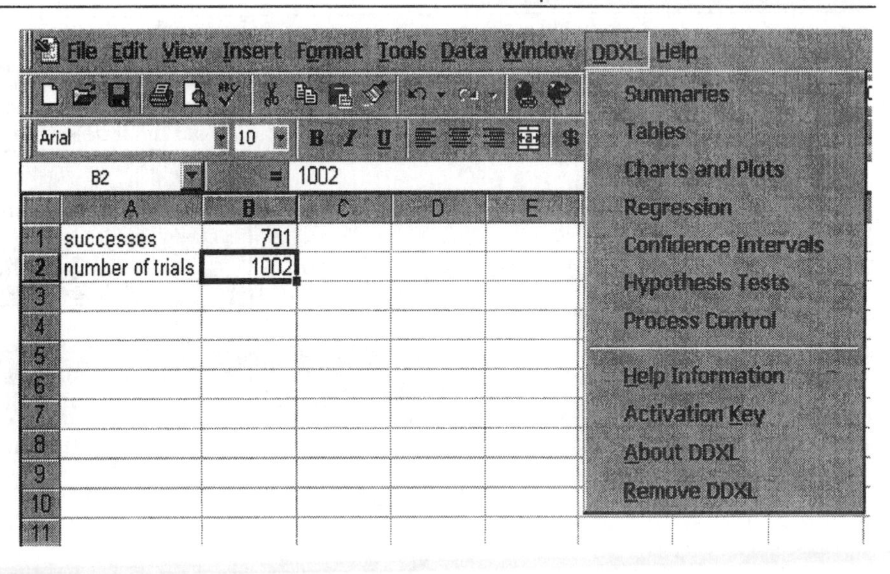

5) Select **Summ 1 Var Prop Interval** as seen below in the screen on the left.

6) Click on the **pencil icon** for **"num successes"** and enter either the cell address B1 or the value 701.

7) Click on the **pencil icon** for **"num trials"** and enter either the cell address B2 or the value 1002.

8) Click **OK.**

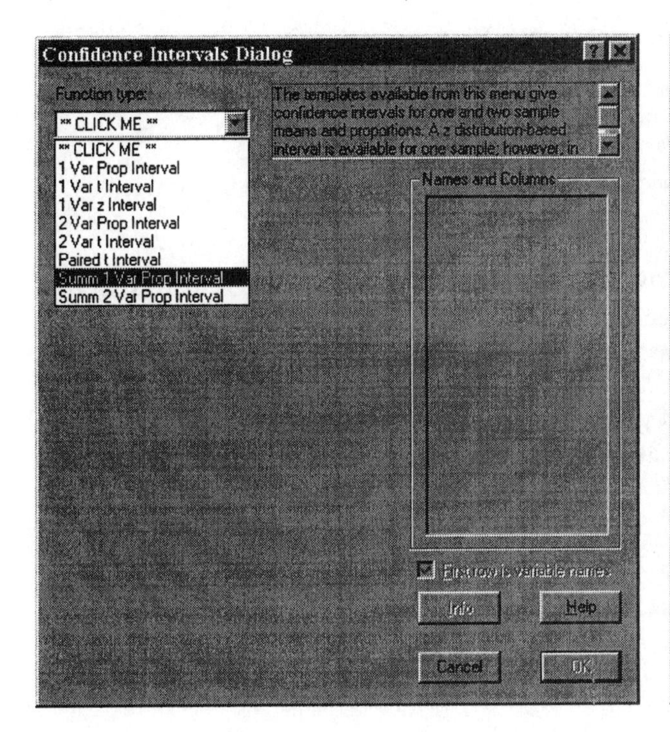

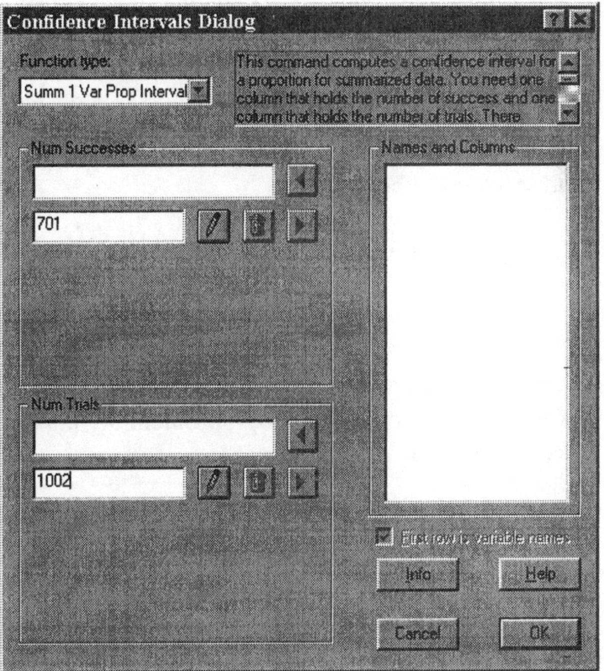

In the dialog box that looks like the one seen on the right

a) Select the appropriate level of confidence, in this case **95%**
b) Click on **Compute Interval**.

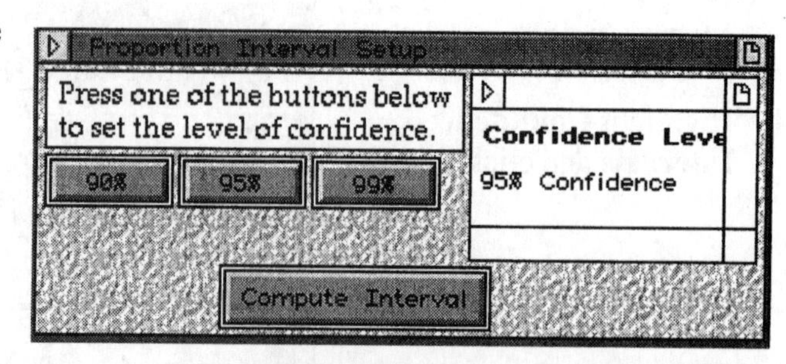

A summary dialog box similar to the one shown on the right displays the confidence interval for the population proportion.

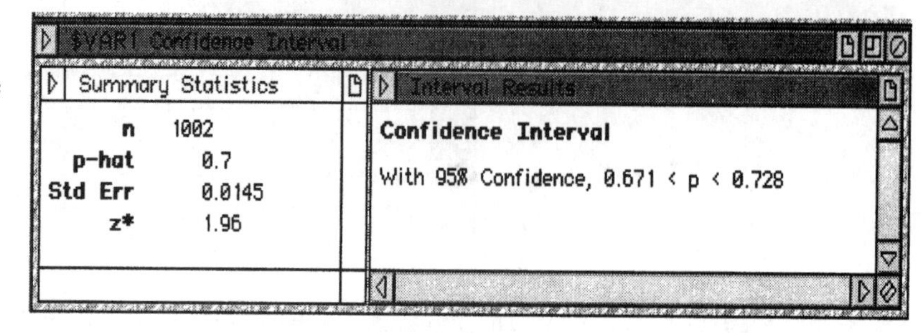

It is possible to **copy and paste your DDXL results** to your Excel worksheet.

1) Click on the title bar of the window you wish to copy into Excel.

2) From the **Edit** menu choose **Copy Window**.

3) Switch to Excel and choose **Paste** from the **Edit** menu or use the **Paste Function** feature.

TO PRACTICE THESE SKILLS

You can practice the technology skills learned in this section by working through the following problems found in your textbook.

1) Use **DDXL** to work through problem 21 on page 339 in the textbook. Note that the problem asks for the point estimate and the confidence interval as a percentage. DDXL will return a decimal value. You will need to rewrite your results in the appropriate format.

2) Use the data found in Data Set 10 in appendix B of your textbook or the data file "M&M" found on the CD data disk to work through problem 33 on page 341. Repeat this problem using brown M&Ms and compare your results.

CHAPTER 7: HYPOTHESIS TESTING

SECTION 7-1: OVERVIEW

This chapter will look at how sample data can be used to make decisions about population parameters. Chapter 6 used sample statistics to estimate population parameters; in this chapter we will use sample statistics to test hypotheses made about population parameters. While Excel has many built in statistical analysis tools available, it does not have a tool for large sample hypothesis tests for the mean. We will use the DDXL add-in for the various hypotheses tests.

SECTION 7-2: TESTING A CLAIM ABOUT A MEAN: LARGE SAMPLES

Begin by opening the Excel file that contains the body temperature data entered at the beginning of the last chapter. This data can also be found in the chapter problem at the beginning of Chapter 7 (Table 7–1 Body Temperatures (F°) of 106 Healthy Adults). We will work through the problem found on the bottom of page 383 in the text which makes use of this data and focuses on testing the hypothesis that the mean body temperature of healthy adults is equal to 98.6°

LARGE SAMPLE TEST FOR μ

1) Select **DDXL** from the tool bar, scroll down to **Hypothesis Tests** and click.
2) From the **Hypothesis Tests Dialog** box select **I Var z Test** from Function type.
3) Click on the pencil icon and list the range of cells that include your data.
4) Click **OK**.
5) Complete each of the four steps listed in the dialog box shown below. These steps are outlined on the next page.

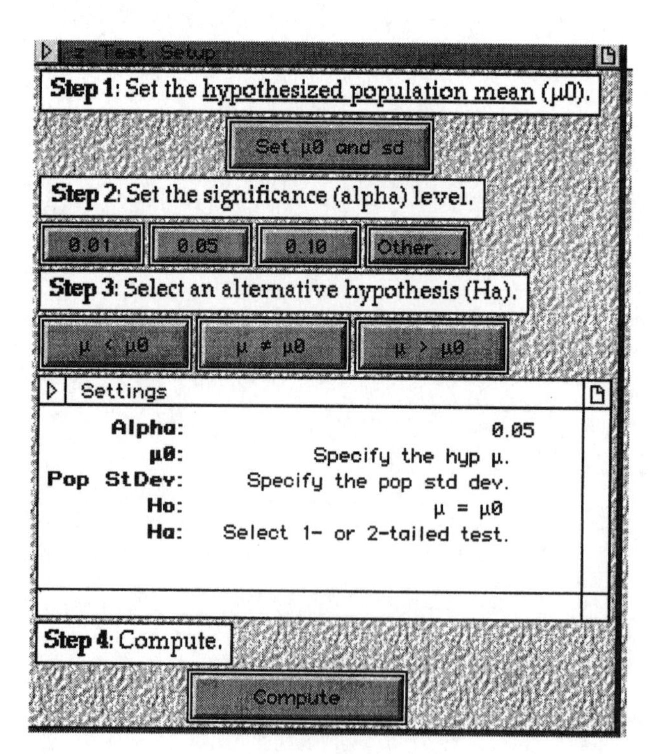

a) **Step 1:** Click on **"Set $\mu\theta$ and sd"** . We found the mean and standard deviation when we worked with this data in the last chapter. If you do not have these values already it is relatively easy to determine the mean and standard deviation using Excel's built in functions. Enter the values for the mean and standard deviation in the appropriate spot and click **OK.**

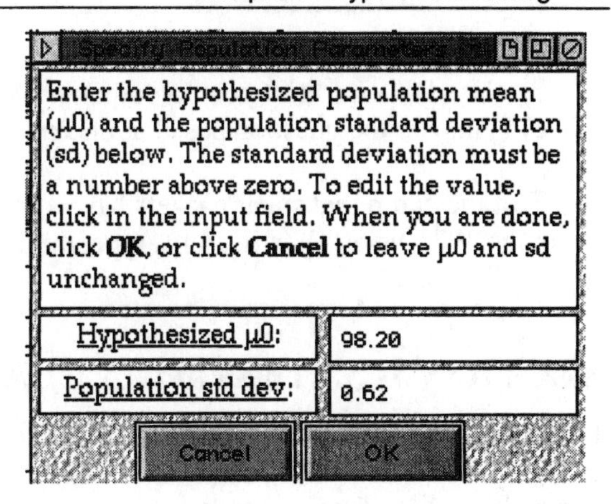

b) **Step 2:** Set the **significance level** by clicking on the appropriate value. In this case use a significance level of 0.05.

c) **Step 3:** Select the **alternative hypothesis**. In this problem the null hypothesis states that the mean body temperature of healthy adults is equal to 98.6°. Therefore the alternative hypothesis states that the mean temperature is not equal to 98.6°. Choose $\mu \neq \mu\theta$.

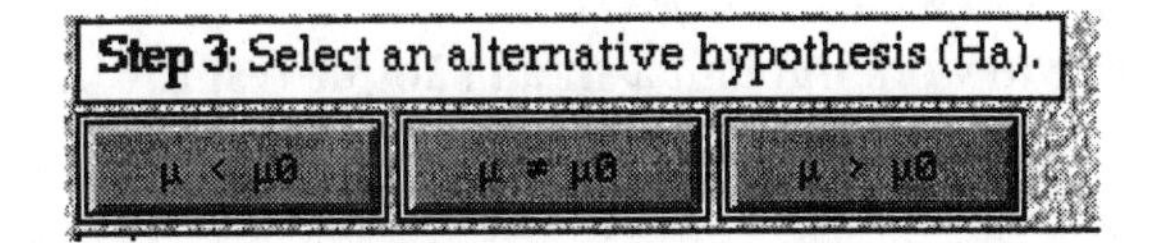

d) **Step 4**: Click on **Compute**.

6) DDXL presents the following results which include the test statistic, the P – value and a conclusion to reject the null hypothesis.

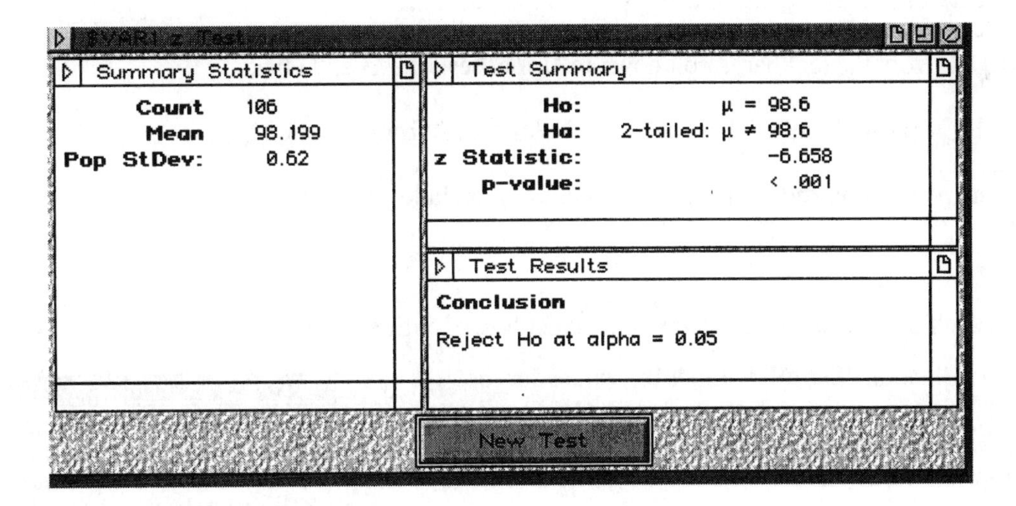

TO PRACTICE THESE SKILLS

You can practice the skills learned in this section by working through the following problems..

1) Using data already saved in an Excel file dealing with the mean body temperature of healthy adults work through problem 9 on page 395.

2) Use the weight of quarters data found in Data Set 13 in Appendix B of your textbook or the data file "QUARTERS" found the CD data disk that accompanies your text to work through problem 19 on page 396.

a) Set up the null hypothesis.
b) Use DDXL to test your hypothesis.
c) Answer the question posed at the end of the problem.

3) Test the claim that sugar packets have a mean equal to 3.5 g as indicated on the label (question 21 on page 397) using Data Set 4 in Appendix B or the file "SUGAR" on the CD data disk.

SECTION 7- 3: TESTING A CLAIM ABOUT A MEAN: SMALL SAMPLES

In this section we turn our attention to small samples, those where the number of sample values is less than or equal to 30, and will look at testing claims made about a population mean when the sample size is small. We will use the *t* distribution that was introduced in the previous chapter dealing with confidence intervals. The criterion for choosing a Student *t* distribution is the same. Excel does not does not have a tool for small sample hypothesis tests for the mean and so we will rely on the DDXL add-in to perform a *t* test of the hypothesis of the mean for a small sample. This test compares the observed *t* test statistic to the point of the *t* distribution that corresponds to the test's chosen α level.
We will use the following example to work through a **small sample test for μ.**

SMALL SAMPLE TEST FOR μ:

Begin by entering the data found in the **Pulse Rates** example found on page 402 into an Excel worksheet. Is there sufficient evidence to support the conclusion that statistics students have an average pulse rate greater than 60 beats per minute? Use a significance level of $\alpha = 0.10$.

1) Begin by entering your data into Excel.

2) Select **DDXL** from the tool bar, scroll down to **Hypothesis Tests** and click.

3) From the **Hypothesis Tests Dialog** box select **I Var t Test** from Function type.

4) Click on the pencil icon and list the range of cells that include your data.

5) Click **OK**.

6) Complete each of the four steps listed in the dialog box as outlined below.

   a) **Step 1:** Click on **"Set μ0 "**. Enter the values for the hypothesized population mean and click **OK**.
   b) **Step 2:** Set the **significance level** by clicking on the appropriate value. In this case use a significance level of 0.01.

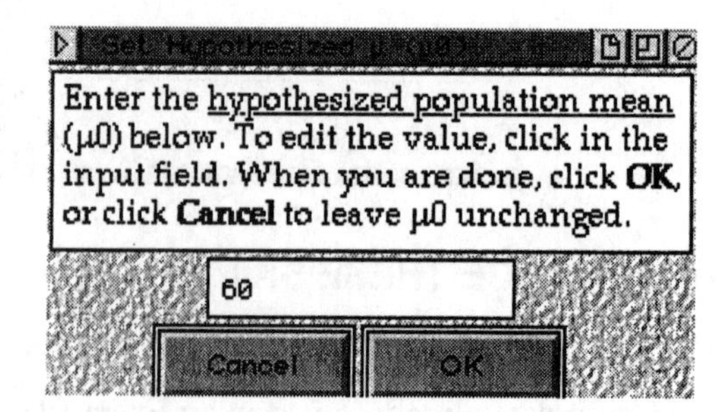

   c) **Step 3:** Select the **alternative hypothesis**. In this problem the null hypothesis states that average pulse rate is less than or equal to 60. The alternative hypothesis states that the average pulse rate is greater than 60. Therefore choose $\mu > \mu 0$.

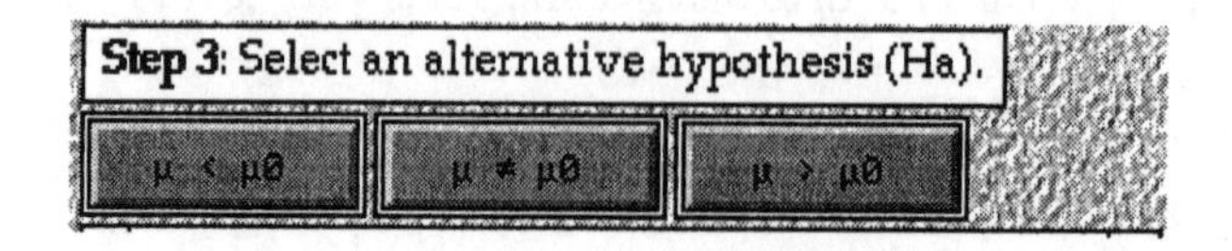

d) **Step 4:** Click on **Compute.**

7) **DDXL** presents the following results which include the test statistic, the P – value and a conclusion to reject the null hypothesis.

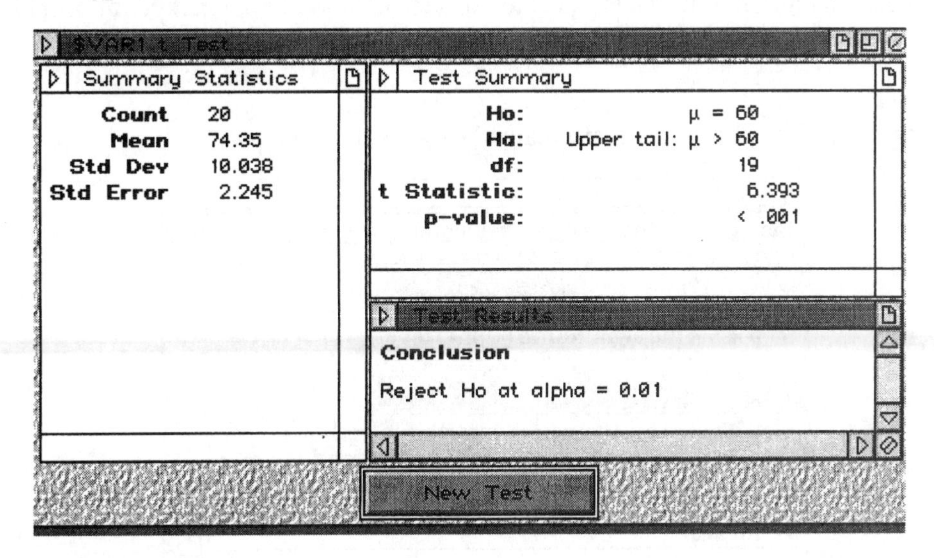

TO PRACTICE THESE SKILLS

You can practice the technology skills learned in this section by working on the following problems.

1) Use the weights of brown M&M candy found in Data Set 10 in Appendix B of your textbook or on the CD data disk that accompanies your text in the file "M&M" to work through problem 17 on page 406 of your textbook.

2) Input the data found in problem 23 on page 407 to test the claim that the mean birth weight for all male babies of mothers given vitamins is equal to 3.39kg.

3) Repeat the preceding problem assuming that the standard normal distribution was incorrectly used instead. How does using the standard normal distribution impact your conclusion? Explain your results.

SECTION 7-4: TESTING A CLAIM ABOUT A PROPORTION

In the last two sections we used Excel and the DDXL add in to test claims made about the population of the mean for both a large and small sample. We will now look at testing claims made about a population proportion. We will utilize the DDXL add in to perform a z test of the hypothesis for a proportion. You will notice a similarity in the basic procedure used to test a claim made about a population.

Z TEST FOR ONE VARIABLE PROPORTION TEST:

Refer to the **Survey of Voters** example found on page 411 of your textbook.

1) Enter the number of people who voted and the number of people surveyed into Excel.

2) Select **DDXL** from the tool bar, scroll down to **Hypothesis Tests** and click.

3) From the **Hypothesis Tests Dialog** box select **Summ 1 Var Prop Test** from Function type.

4) Click on the pencil icon for **Num Successes** and enter the cell address for the voters. Be sure to enter the cell address rather than the actual number of number of people who voted.

5) Click on the pencil icon for **Num Trials** and enter the cell address of total number of people surveyed.

6) Click **OK**.

7) Complete each of the four steps listed in the dialog box as outlined below.

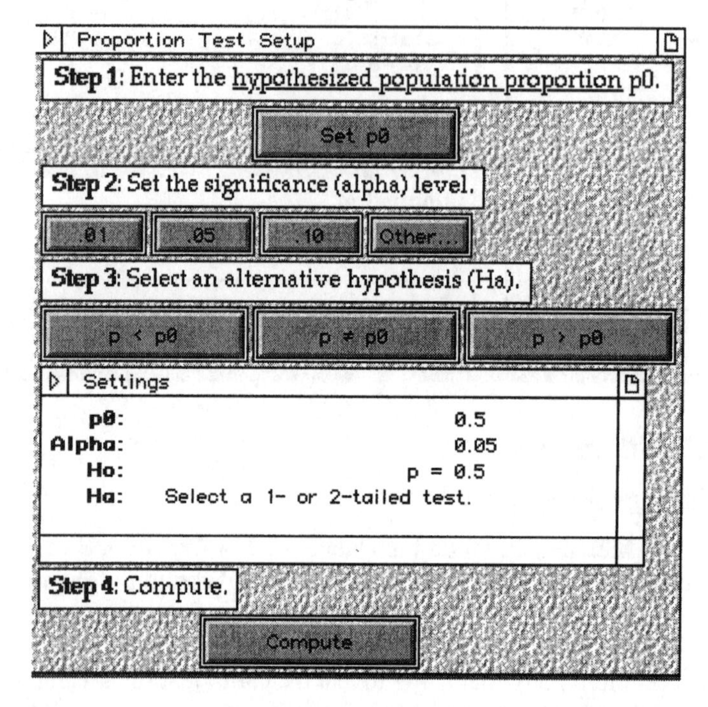

a) **Step 1:** Click on **"Set pθ "**. We will enter the hypothesized test proportion found within the example ($p = 0.61$) Enter the value in the appropriate spot and click **OK.**
b) **Step 2:** Set the **significance level** by clicking on the appropriate value. In this case use a significance level of 0.05.
c) **Step 3**: Select the alternative hypothesis. In this problem the null hypothesis states that the proportion of people who actually did vote was equal to 0.61. Therefore the alternative hypothesis states that the proportion of people who did vote is not equal to 0.61. Choose $p \neq p\theta$.
d) **Step 4**: Click on **Compute.**

8) **DDXL** presents the following results which include the test statistic, the P – value and a conclusion to reject the null hypothesis.

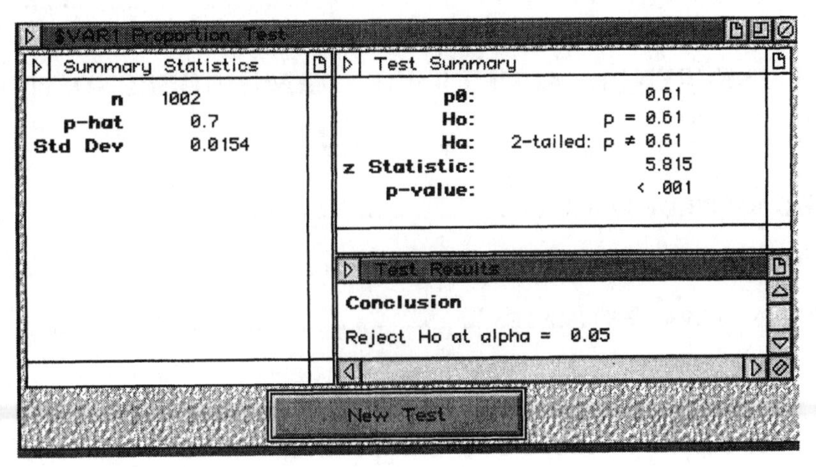

TO PRACTICE THESE SKILLS

You can apply the technology skills learned in this section by working on the following problems.

1) Use the information given in problem 7 on page 414 to test the claim that with scanners 1% of sales are overcharges.

2) From the Review Exercises found at the end of the chapter work on problem 8 on page 428.

CHAPTER 8: INFERENCES FROM TWO SAMPLES

SECTION 8-1: OVERVIEW

In Chapters 6 and 7 we looked at confidence intervals and hypothesis testing as it applied to a single sample. We now turn our attention to confidence intervals and hypothesis tests for comparing two samples. If sample data is available and the sample size is large, $n \geq 30$, to test the hypothesis made about two population means we can use.

- the Data Analysis **z-Test: Two Samples for Means.**
- the DDXL add-in Hypothesis Tests: **2 Var t Test.**

z-Test: Two Samples for Means
This analysis tool is used to perform a two-sample z-test for means with known variances. It is also used to test hypotheses about the difference between two population means.
If sample data is available and the sample size is large, $n \geq 30$, to test the hypothesis made about two population variances we can use.

F Test Two Samples for Variances
This analysis tool performs a two-sample F-test to compare two population variances and to determine whether the two population variances are equal. It returns the *p* value of the one tailed F statistic, based on the hypothesis that array 1 and array 2 have the same variance.

If sample data is available and the sample size is $n \leq 30$, to test the hypothesis made about two population variances we can use.

t-Test: Two Samples Assuming Equal Variances
This test calculates a two sample Student *t* Test. The test assumes that the variance in each of the two groups is equal. The output includes both one tailed and two tailed critical values.

t-Test: Two Samples Assuming Unequal Variances.
This test calculates a two sample Student *t* Test. The test allows the variances in the two groups to be unequal. The output includes both one tailed and two tailed critical values.

SECTION 8-2: INFERENCES ABOUT TWO MEANS: INDEPENDENT AND LARGE SAMPLES

In this section we will work with sample data containing two independent samples with size $n \geq 30$ to test the hypothesis made about two population means. If sample data is available we can use

- the Data Analysis **z-Test: Two Samples for Means.**
- the DDXL add-in Hypothesis Tests: **2 Var t Test.**

Both methods will be outlined.

We will look at the **Coke versus Pepsi** problem found on page 440 of the textbook. The data for this problem can be found in Data Set 1 in Appendix B of your textbook or on the data CD in the file "COLA". Use the weights of samples of regular Coke and the weights of samples of regular Pepsi in the Excel spreadsheet.

Note:
It is suggested that you save this data as it will be used in several sections through out this chapter.

Z-TEST: TWO SAMPLE FOR MEANS

1) Determine the mean (AVERAGE), variance (VAR) and standard deviation (STDEV) of both samples using the built in functions found in Excel.

   A summary of that information should look like the information presented here. A similar table can be found on page 440 in the text.

	regular coke	regular pepsi
mean	0.81682	0.82410
variance	0.000056	0.000033
standard deviation	0.007507	0.005701

2) Click on **Tools,** highlight **Data Analysis** and click.

3) From the Analysis Tools list box in the Data Analysis dialog box select **z-Test: Two Sample for Means.**

4) In the **z-test: Two Sample for Means** dialog box enter the following information:
   a) Weights for regular Coke in the **Variable 1 Range** box.
   b) Weights for regular Pepsi in the **Variable 2 Range** box.
   c) You can enter 0 in the **Hypothesized Mean Difference** box or just leave it blank.
   d) Enter the variance for the Coke in **Variable 1 Variance** and the variance for Pepsi in the **Variable 2 Variance** box.
   e) Enter 0.01 in the **Alpha** box. This value is supplied in the problem.
   f) Determine where you wish to display the output.
   g) Click **OK.**

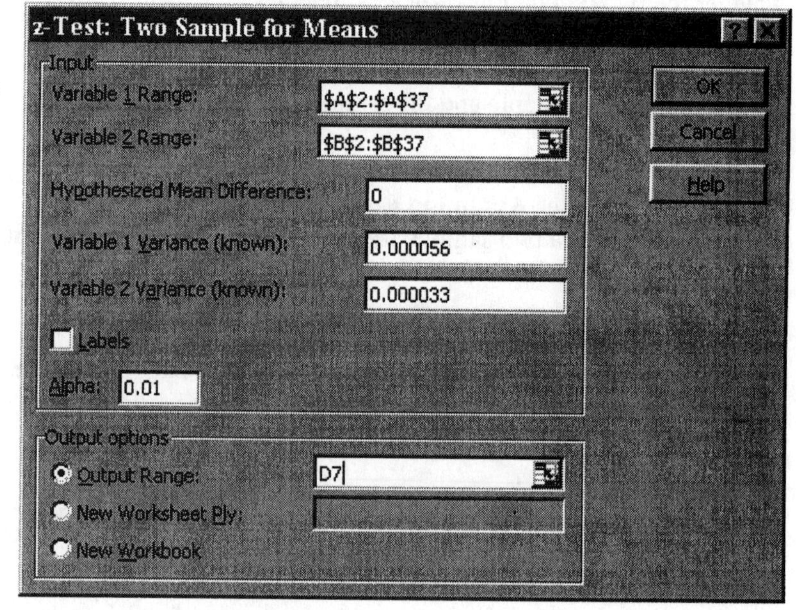

5) The following summary of information containing calculations for the z-test is in added to your current Excel worksheet.

z-Test: Two Sample for Means		
	Variable 1	*Variable 2*
Mean	0.816822222	0.8241028
Known Variance	0.000056	0.000033
Observations	36	36
Hypothesized Mean Difference	0	
z	-4.630424072	
P(Z<=z) one-tail	1.82655E-06	
z Critical one-tail	2.326341928	
P(Z<=z) two-tail	3.65309E-06	
z Critical two-tail	2.575834515	

Note:
It is worth mentioning that this chart is not a "live" chart so that any changes made to the original data at this point would require using the data analysis tool a second time to produce new test results.

DDXL – 2 VAR T TEST

Copy the weights of samples of regular Coke and the weights of samples of regular Pepsi to a new worksheet.

1) Click on **DDXL**, select **Hypothesis Tests** and **2 Var t Test.**

2) In the dialog box click on the pencil icon for the **1st Quantitative Variable** and enter the range of data for the weight of regular Coke as you did before. Then click on the pencil icon for the **2nd Quantitative Variable** and enter the range of data for the weight of regular Pepsi. Click OK

3) Follow these steps in the **2 Sample t Test Setup** dialog box:

   a) **Step 1:** Select **2-sample.**
   b) **Step 2:** This step is optional so you can skip over it or set the difference at 0.
   c) **Step 3:** Set the significance level at 0.01.
   d) **Step 4:** Select the appropriate alternative hypothesis.
   e) **Step 5:** Click on **Compute.**

4) The results can be seen below. In addition to the Test Summary, DDXL also returns the mean and standard deviation of each variable and a conclusion to reject the null hypothesis.

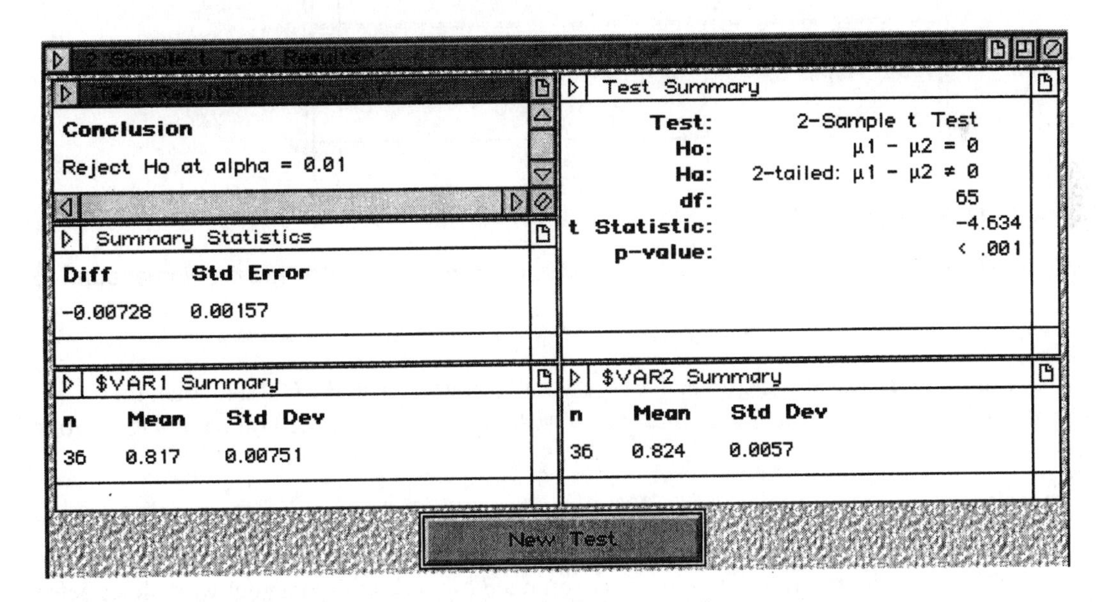

CONFIDENCE INTERVAL ESTIMATES

Using the sample data for **Coke and Pepsi** we can construct a confidence interval estimate of the difference between the mean weight of regular Coke and regular Pepsi. This will be done using the DDXL add-in. The procedure for doing this is very similar to the one used to determine the 2 Var t Test.

1) Click on **DDXL**, select **Confidence Interval – 2 Var t Test.**

2) In the dialog box click on the pencil icon for the 1st **Quantitative Variable** and enter the range of data for the weight of regular Coke as you did before. Then click on the pencil icon for the 2nd **Quantitative Variable** and enter the range of data for the weight of regular Pepsi. Click **OK.**

3) In **Step 1** choose **2 sample**. In **Step 2** select the appropriate confidence level (in this case 99%) and in **Step 3** click on **Compute Interval.**

4) The following results are displayed. In addition to the Confidence Interval results DDXL also returns the mean and standard deviation for each variable.

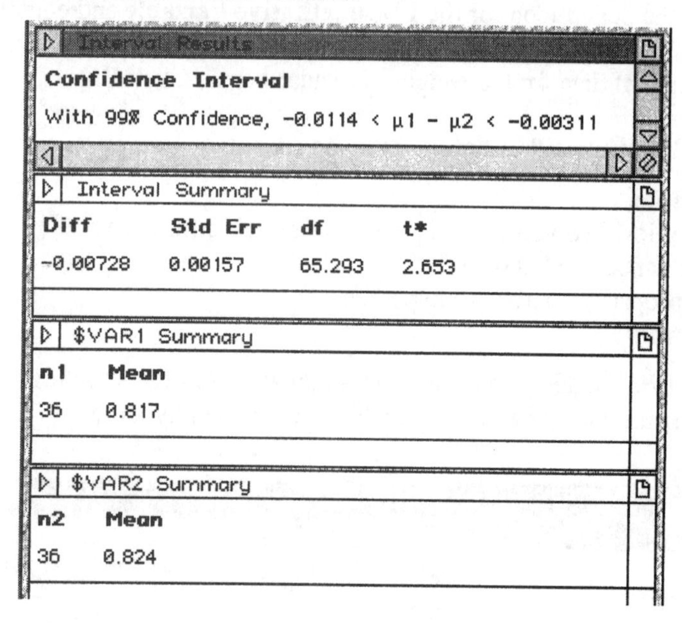

TO PRACTICE THESE SKILLS

You can practice the skills learned in this section by working through the following problems found in your textbook.

1) Use the data found in Data Set 2 in Appendix B of your textbook or on the CD data disk that accompanies your text in the file "TEXTBOOK" to work through problem 13 on page 446 of your textbook.

2) Problem 16 on page 446-447 tests the claim that male and female statistics students have the same mean heart rate. Use the data from Data Set 14 in Appendix B or on the CD data disk file "STATSURV". You may find it easier to copy just those columns you need to work with to a new worksheet before beginning this exercise.

SECTION 8-3: INFERENCES ABOUT TWO MEANS: MATCHED PAIRS

We now turn our attention to confidence intervals and hypothesis tests for comparing two samples. The population parameters include the difference between two population means $\mu_1 - \mu_2$ using paired observations (matched pairs). The analysis can be done using either

- the Data Analysis **t-Test: PairedTwo Samples for Means.**
- the DDXL add-in Hypothesis Tests: **Paired Test**

Begin by entering the data (Table 8-1 Reported and Measured Heights of Male Statistics Students) found on the top of page 449 in your textbook into an Excel spreadsheet. Enter the data reported height, measured height and differences in columns A, B and C respectively. The Data Analysis t Test requires that the data for each group be in a separate column. This is often referred to as **unstacked data.**

As we examine the data we have entered into Excel we notice that one of the values listed in the differences column is noticeably higher than any of the other differences. This student's reported height appears to be an **outlier**. We will remove this student from our data before we begin our statistical analysis.

T-TEST: PAIRED TWO SAMPLES FOR MEANS

Using the data entered into Excel, we will test the claim made in the example found on page 451 in your textbook that male statistics students do exaggerate by reporting heights that are greater than their actual measured heights,

1) Click on **Tools,** highlight **Data Analysis** and click.

2) From the Analysis Tools list box in the Data Analysis dialog box select **t-Test: Paired Two Samples for Means.** Click **OK**.

3) In the t Test: Paired Two Samples for Means dialog box enter the following information
   a) Cell range containing reported height in the **Variable 1 Range** box.
   b) Cell range for measured height in the **Variable 2 Range** box.
   c) Enter 0 in the Hypothesized Mean Difference box.
   d) Enter 0.05 in the **Alpha** box. This value is supplied in the problem.
   e) Determine where you wish to display the output.
   f) Click **OK**.

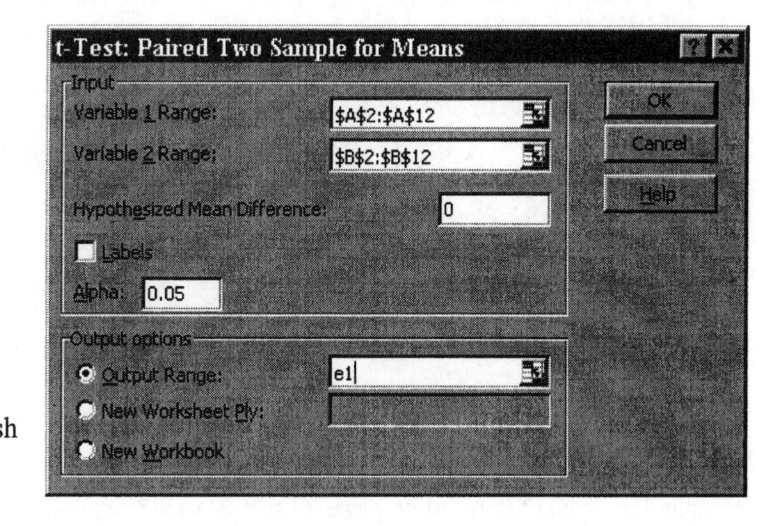

4) The following summary of information containing calculations for the t-test is in added to your current Excel worksheet.

t-Test: Paired Two Sample for Means		
	Variable 1	*Variable 2*
Mean	69.22727273	68.55454545
Variance	4.468181818	4.386727273
Observations	11	11
Pearson Correlation	0.922999095	
Hypothesized Mean Difference	0	
df	10	
t Stat	2.701377679	
P(T<=t) one-tail	0.011130329	
t Critical one-tail	1.812461505	
P(T<=t) two-tail	0.022260659	
t Critical two-tail	2.228139238	

In addition to the test statistic, Excel displays the P values for a one and two tailed test as well as the corresponding critical values.

DDXL – PAIRED T TEST

1) Click on **DDXL** and choose **Hypotheses Tests** and **Pair t Test.**

2) In the Hypothesis test dialog box click on the pencil icon for the 1st **Quantitative Variable** and enter the range of data for the reported height. Then click on the pencil icon for the 2nd **Quantitative Variable** and enter the range of data for the measured height.

3) Click **OK**.

4) In a manner very similar to the previous hypotheses tests done with DDXL complete the fours steps in the **paired t test** dialog box. Set the **significance level** at 0.05 and select **μ(diff) > μ(diff)θ.**

5) Click on **Compute.**

6) The following results will be displayed.

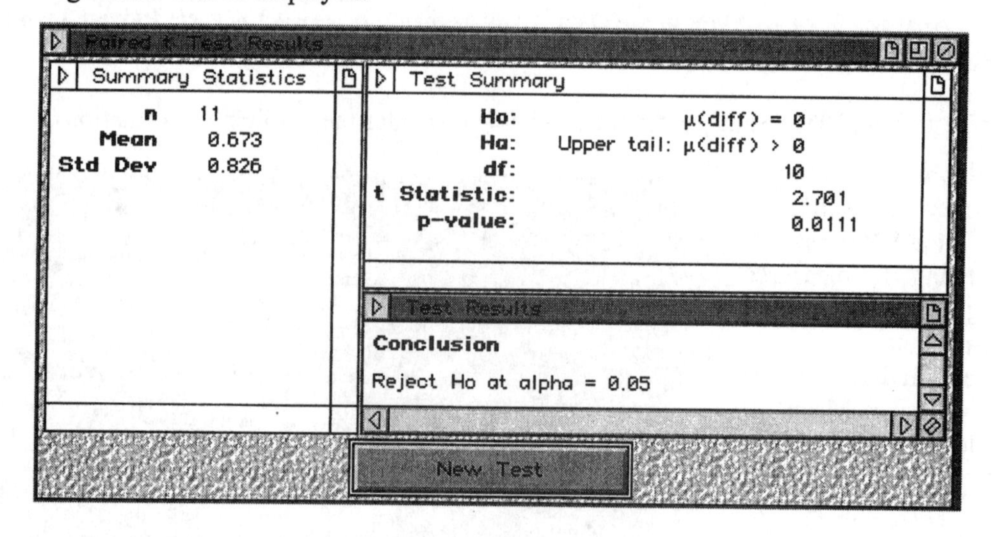

DDXL – CONFIDENCE INTERVALS FOR MATCHED PAIRS

Using the sample data from the preceding pages (Table 8–1) we can construct a confidence interval estimate of the mean of the differences between the reported heights and measured heights of male statistic students. This is the focus of the example found on the top of page 453 of your textbook. This will be done using the DDXL add-in. The procedure for doing this is very similar to the one used to determine the 2 Var t Test.

1) Click on **DDXL**, select **Confidence Interval – Paired t Interval.**

2) In the dialog box click on the pencil icon for the 1st **Quantitative Variable** and enter the range of data for the reported height. Then click on the pencil icon for the 2nd **Quantitative Variable** and enter the range of data for the measured height.

3) Click **OK.**

4) Select the appropriate confidence level (in this case 95%) and click on **Compute Interval.**

5) The following results are displayed.

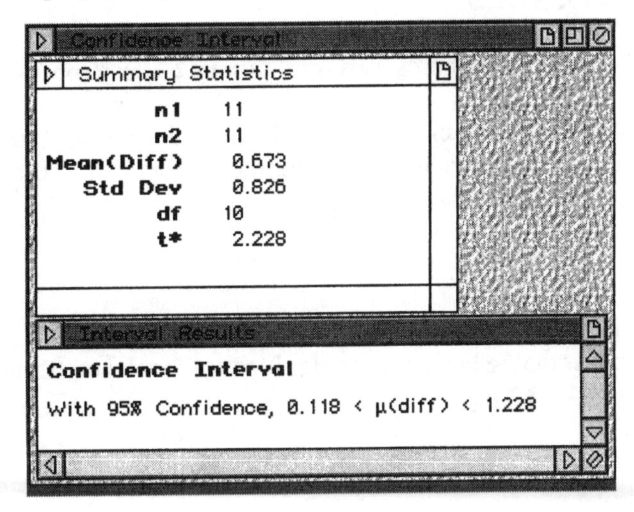

TO PRACTICE THESE SKILLS

You can practice the technology skills learned in this section by working on the following problems.

1) Use the data presented in problem 7 on page 455 that lists SAT scores before and after students took a preparatory course to conclude whether or not a preparatory course does in fact improve scores. Construct the confidence interval indicated and interpret the results.

2) Work through problem 2 in the Review Exercises for Chapter 8, found on page 494.

SECTION 8-4: INFERENCES ABOUT TWO PROPORTIONS

To perform the Z test for the differences in two proportions use DDXL add-in. This will require that we enter the number of successes and trials for Sample 1 and the number of successes and trails for Sample 2 into Excel.

Using the information presented to us in the **Viagra Treatment and Placebo** example found on the bottom of page 460 of your textbook we note

- Viagra – 734 trials with 16% experiencing headaches.
- Placebo – 725 trials with 4% experiencing headaches.

In entering this information into Excel we produce the following:

	A	B	C
1		successes	trials
2	viagra	117	734
3	placebo	29	725

DDXL – SUMM 2 VAR PROP TEST

1) Click on **DDXL** and choose **Hypotheses Tests** and **Summ 2 Var Prop Test.**

2) In the Hypothesis test dialog box click on the pencil icon and enter the number of successes for Viagra followed by the number of Viagra trials. Repeat for the number of successes for the placebo group and the number of trials conducted for this group.

3) Click **OK.**

4) In a manner very similar to the previous hypotheses tests done with DDXL complete the fours steps in the **summ 2 var prop test** dialog box. Set the **significance level** at 0.01 and select the alternative hypothesis, in this case **p1 – p2 > p.**

5) Click on **Compute.**

6) The following information will be displayed.

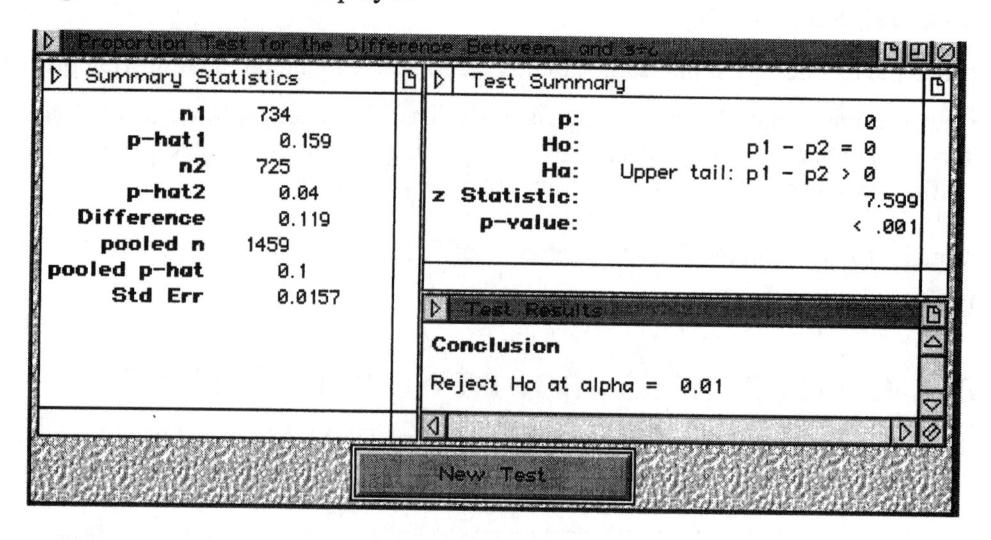

DDXL - CONFIDENCE INTERVALS FOR PROPORTION PAIRS

Using the sample data from the preceding example we can construct a confidence interval estimate of the difference between population proportions using DDXL. The method for doing this is very similar to the one used to determine the Paired t Interval.

1) Click on **DDXL**, select **Confidence Interval – Summ 2 Var Prop Interval.**

2) In the dialog box click on the pencil icon and enter the number of successes for Viagra followed by the number of Viagra trials. Repeat for the number of successes for the placebo group and the number of trials conducted for this group.

3) Click **OK.**

4) Select the appropriate confidence level (in this case 99%) and click on **Compute Interval.**

5) The following results are displayed.

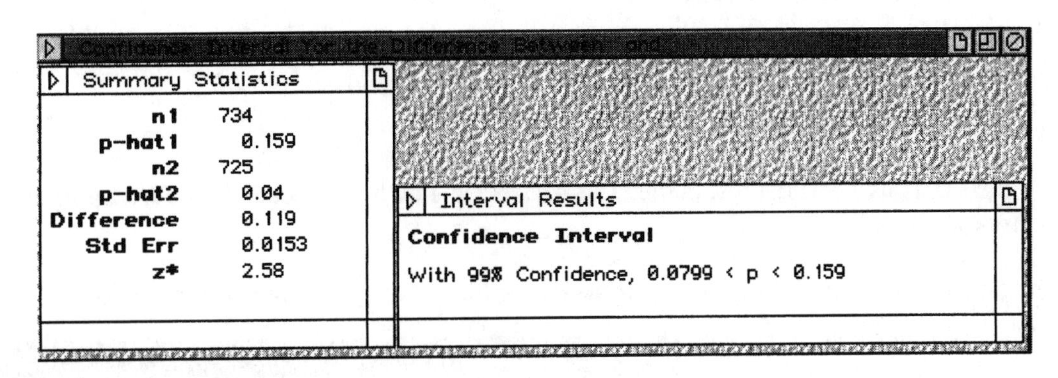

TO PRACTICE THESE SKILLS

You can practice the skills covered in this section by working on the following problems found in your textbook.

1) Problem 9 on page 466 focuses on the use of seat belts and hospital time for children. Work through this problem using the motes outlined in this section.

2) Problem 15 on page 467-468 asks you to compare the results of a student survey given in written form and via computer. Pay close attention to the questions asked following this problem.

3) Problem 22 on page 469 is excellent practice in comparing three independent populations. If you are feeling adventurous you may want to try this problem.

SECTION 8.5: COMPARING VARIATION IN TWO SAMPLES

In this section we present a test of hypothesis for comparing two population variances. This will be done using the **F Test Two Sample for Variances** found in Excel's Data Analysis tools. This test can only be used when sample data is available. One thing to be aware of is that this test is done as an upper tail test. So to test whether the variances are not equal, you will need to enter half the value of α that you want to use and you must enter the larger various as population 1.

Earlier in this chapter we determined these values using the weights of samples of regular Coke and regular Pepsi.

	regular coke	regular pepsi
mean	0.81682	0.82410
variance	0.000056	0.000033
standard deviation	0.007507	0.005701

Using this information and mirroring the problem introduced in the example found on page 473 in your textbook we will use the **F Test Two Sample for Variance** to test the stated hypothesis.

F-TEST: TWO SAMPLE FOR VARIANCES

Using the data entered into Excel, we will test the claim that the weights of samples of regular Coke and the weights of samples of regular Pepsi have the same standard deviation.

1) Click on **Tools,** highlight **Data Analysis** and click.

2) From the Analysis Tools list box in the Data Analysis dialog box select **F-Test Two Sample for Variances.** Click **OK**.

3) In the t Test: Paired Two Samples for Means dialog box enter the following information:

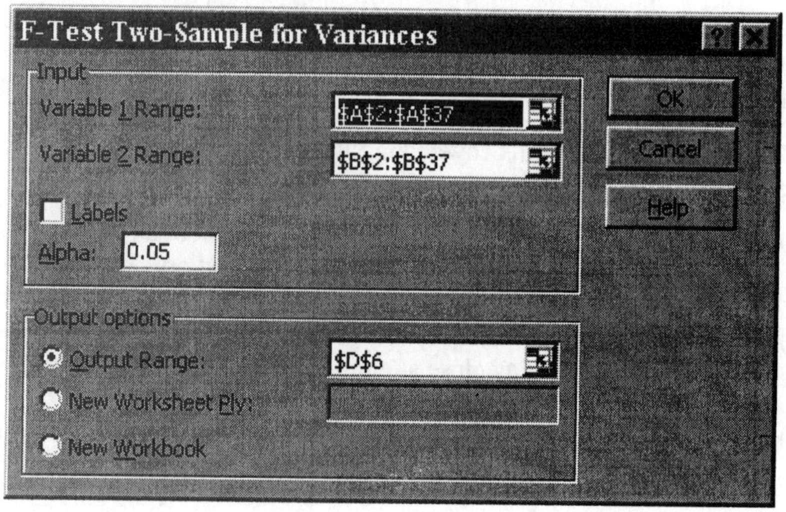

   a) The cell range containing the weight of regular Coke samples in the **Variable 1 Range** box.
   b) The cell range for the weight of regular Pepsi in the **Variable 2 Range** box.
   c) Since the range of cells in (a) and (b) did not include the labels for the columns leave this box unchecked. If you did include the labels check this box.
   d) Enter 0.05 in the **Alpha** box. This value is supplied in the problem.
   e) Determine where you wish to display the output.
   f) Click **OK**.

4) The following summary of information containing calculations for the F-test is in added to your current Excel worksheet.

F-Test Two-Sample for Variances

	Variable 1	Variable 2
Mean	0.817	0.824
Variance	0.000	0.000
Observations	36.000	36.000
df	35.000	35.000
F	1.734	
P(F<=f) one-tail	0.054	
F Critical one-tail	1.757	

In addition to the critical value of F as the critical value of F this summary also contains the P-value for the one tailed case. This value can be doubled for a two tailed test.

TO PRACTICE THESE SKILLS

You can practice the technology skills learned in this section by working on the following problems found in your textbook.

1) Problem 9 on page 477 compares "test anxiety" scores of students when test questions are arrange in order of difficulty, i.e., from easiest to most difficulty and from most difficult to easiest.
    a) Enter the data presented in the problem into Excel.
    b) Compare your results with those presented on page 477.
    c) Test the claim that the two samples come from populations with the same variance.

Save this data as we will use it again in the following section.

SECTION 8-6: INFERENCES ABOUT TWO MEANS: INDEPENDENT AND SMALL SAMPLES

In this section we will examine hypotheses tests and confidence intervals involving the means of two independent samples when at least on of the sample sizes is small ($n \leq 30$). If sample data is available we can use

- the Data Analysis **t-Test: Two Samples Assuming Equal Variances**
  t-Test: Two Samples Assuming Unequal Variances.
- the DDXL Add-In Hypothesis Tests **2 Var t Test**
  Confidence Interval **2 Var t Test**

Begin by entering the data found in Table 8-2 (Nicotine Contents of Cigarettes) on page 481 into an Excel worksheet.

T-TEST: TWO SAMPLES ASSUMING EQUAL VARIANCES

The process is very similar to methods used throughout this chapter. We will test the claim that the mean amount of nicotine in filtered cigarettes is equal to the mean amount of nicotine in unfiltered cigarettes.

1) Click on **Tools,** highlight **Data Analysis** and click.

2) From the Analysis Tools list box in the Data Analysis dialog box select **t-Test: Two Sample Assuming Equal Variances.** Click **OK**.

3) In the t Test: Two Sample Assuming Equal Variances dialog box enter the following information.
    a) Enter the cell range for filtered cigarettes in the **Variable 1 Range** box.
    b) Enter the cell range for non-filtered cigarettes in the **Variable 2 Range** box.
    c) Enter 0 in the **Hypothesized Mean Difference** box.
    d) Since the range of cells did not include the labels for the columns, leave this box unchecked. If you did include the labels, check this box.

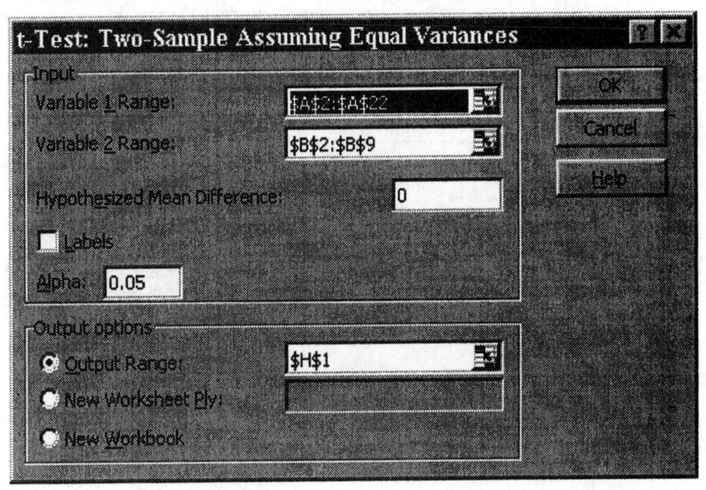

e) Enter 0.05 in the **Alpha** box. This value is supplied in the problem.
f) Determine where you wish to display the output.
g) Click **OK**.

4) The following summary of information containing calculations for the t-test is in added to your current Excel worksheet.

t-Test: Two-Sample Assuming Equal Variances		
	filtered	*non-filtered*
Mean	0.9428571	1.6375
Variance	0.0955714	0.028392857
Observations	21	8
Pooled Variance	0.0781548	
Hypothesized Mean	0	
df	27	
t Stat	-5.9805275	
P(T<=t) one-tail	1.114E-06	
t Critical one-tail	1.703288	
P(T<=t) two-tail	2.227E-06	
t Critical two-tail	2.0518291	

5) The test statistic falls in the critical region suggesting that we reject the null hypothesis.

DDXL – 2 VAR T TEST

Copy the data from columns A and B to a new worksheet.

1) Click on DDXL, select Hypothesis Tests and **2 Var t Test.**

2) In the dialog box click on the pencil icon for the 1st **Quantitative Variable** and enter the range of data for the filtered cigarettes. Then click on the pencil icon for the 2nd **Quantitative Variable** and enter the range of data for the non-filtered cigarettes. Click **OK.**

3) Follow these steps in the **2 Sample t Test Setup** dialog box:

a) **Step 1:** Click on **Pooled.**
b) **Step 2:** This step is optional so you can skip over it or set the difference at 0.
c) **Step 3:** Set the significance level at 0.05.
d) **Step 4:** Select the appropriate alternative hypothesis.
e) **Step 5:** Click on **Compute.**

4) The following results are displayed. In addition to the Test Summary, DDXL also returns the mean and standard deviation of each variable and a conclusion to reject the null hypothesis.

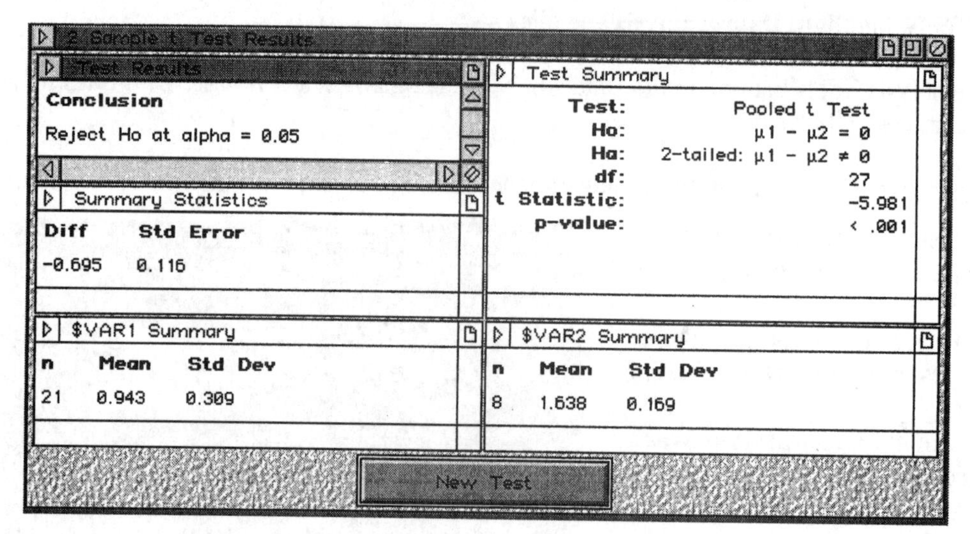

T-TEST: TWO SAMPLES ASSUMING UNEQUAL VARIANCES

Begin by entering the data found in Table 8-3 (Tar Contents of Cigarettes) on page 485 of your textbook into an Excel worksheet.

The process for working with this t-Test is very similar to methods already used throughout this chapter. We will test the claim that the mean amount of tar in filtered cigarettes is less than the mean amount of tar in unfiltered cigarettes.

1) Click on **Tools,** highlight **Data Analysis** and click.

2) From the Analysis Tools list box in the Data Analysis dialog box select **t-Test: Two Sample Assuming Unequal Variances.** Click **OK**.

3) In the t Test: Two Sample Assuming unequal Variances dialog box enter the following information
   a) Enter the cell range for filtered cigarettes in the **Variable 1 Range** box.
   b) Enter the cell range for non-filtered cigarettes in the **Variable 2 Range** box.
   c) Enter 0 in the **Hypothesized Mean Difference** box.
   d) Since the range of cells in (a) and (b) did not include the labels for the columns leave this box unchecked. If you did include the labels check this box.
   e) Enter 0.05 in the **Alpha** box. This value is supplied in the problem.
   f) Determine where you wish to display the output.
   g) Click **OK**.

4) The following summary of information containing calculations for the t-test is in added to your current Excel worksheet.

t-Test: Two-Sample Assuming Unequal Variances		
	Variable 1	*Variable 2*
Mean	13.285714	24
Variance	14.014286	2.857142857
Observations	21	8
Hypothesized Mean Difference	0	
df	26	
t Stat	-10.585452	
P(T<=t) one-tail	3.2E-11	
t Critical one-tail	1.7056163	
P(T<=t) two-tail	6.4E-11	
t Critical two-tail	2.0555308	

CONFIDENCE INTERVAL

Using the sample data for **Nicotine Content (Table 8-2)** we can construct a confidence interval estimate. The methodology outlined here is the same for equal population variances and unequal population variances. This process involves using DDXL and is almost identical to the method outlined in section 8-2 used to determine the 2 Var t Test.

1) Click on **DDXL**, select **Confidence Interval – 2 Var t Test.**

2) In the dialog box click on the pencil icon for the **1st Quantitative Variable** and enter the range of data for the filtered cigarettes as you did before. Then click on the pencil icon for the **2nd Quantitative Variable** and enter the range of data for unfiltered cigarettes. Click **OK**.

3) In **Step 1** choose **pooled** if assuming equal population variances and choose **2 sample** if assuming unequal population variances.

4) In **Step 2** select the appropriate confidence level (in this case 95%) and in **Step 3** click on **Compute Interval.**

5) The following results are displayed. In addition to the Confidence Interval results DDXL also returns the mean and standard deviation for each variable.

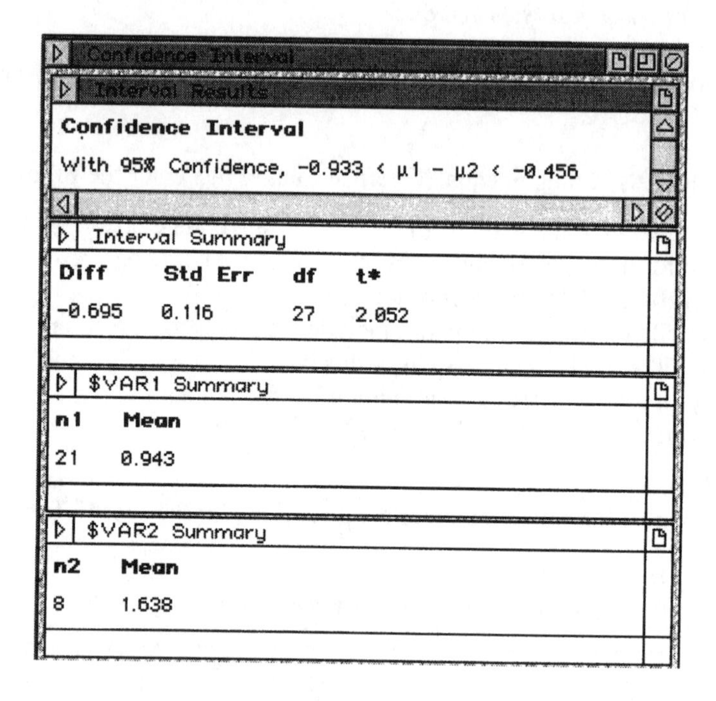

TO PRACTICE THESE SKILLS

You can practice the technology skills learned in this section by working on the following problems.

1) Use the information found in problem 13 on page 490-491 to claim that the mean amount of carbon monoxide is the same for filtered and non-filtered king size cigarettes.

2) Problem 16 on page 477 compares "test anxiety" scores of students when test questions are arrange in order of difficulty, i.e., from easiest to most difficulty and from most difficult to easiest. This data was entered into Excel in the preceding section. Test the claim that order does not impact the mean scores of the two populations

CHAPTER 9: CORRELATION AND REGRESSION

SECTION 9-1: OVERVIEW

In this chapter we will be working with data that comes in pairs. We will be determining whether there is a relationship between the paired data, and will be trying to identify the relationship if it exists.

Excel provides an excellent tool to help us consider whether there is a statistically significant relationship between two variables. We can create a scatter plot, find a line of regression, and use our regression equation to predict values for one of the variables when we know values of the other variable.

The new functions introduced in this section are outlined below.

CORREL

This returns the correlation coefficient between two data sets. Your paired data must be entered in adjacent columns.

ADD TRENDLINE

This feature adds the linear regression graph to the scatter plot of a set of data values.

REGRESSION

This function returns information on Regression Statistics, as well as other information based on the linear regression equation. Your data values must be entered in adjacent columns.

SECTION 9-2: CORRELATION

In this section, we will take a look at a picture of a collection of paired sample data to help us determine if there appears to be a relationship between the two variables. We will work with the data from the chapter problem to learn how to create a scatter plot for the ordered pairs.

1) In cell A1, type in "Bill". In cell B1, type in "Tip".

2) Enter the paired data in column A and B, making sure that you keep the pairs as they are listed. To have the values show up as currency, you may want to format your cells, selecting currency with 2 decimal places.

3) In cell C1, type" r =", and move your cursor to cell D1. To find the linear correlation coefficient, click on the **Function** icon on the main menu bar, or click on **Insert, Function.** In the dialog box that opens, click on **Statistical, CORREL.** Then click on **OK.**

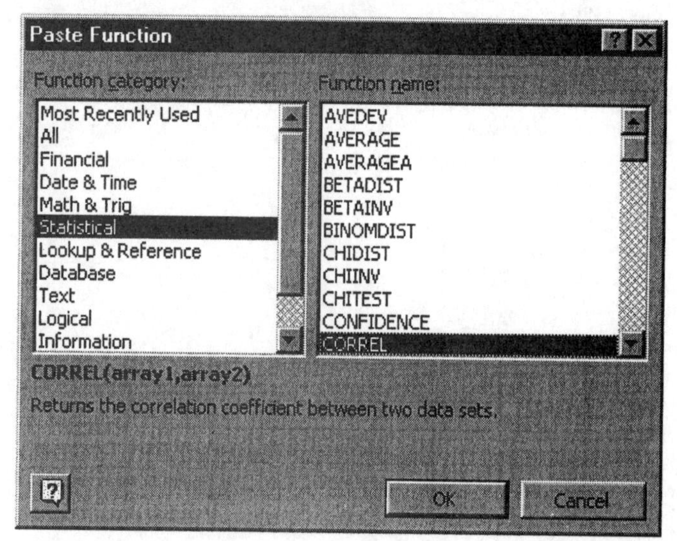

4) In **Array1**, enter the range of cells where the data on the amount of the bill is stored. In **Array2**, enter the range of cells where the data on the tip amount is stored. Then click on **OK**. You will see the correlation coefficient in cell D1. This value should be 0.828159. Read how we can interpret this value in the Example starting on page 511 of your textbook.

5) To create the scatterplot for this data, click on the **Chart Wizard** icon on the main menu, or click on **Insert, Chart.** Click on **XY(Scatter)**, and then click on **Next**.

6) In the **Data Range** box, enter the range of cells where your paired data is stored. If you entered your data in columns A and B starting with cell A2, you could type in "A2:B7". Make sure that the bullet by **Columns** is marked. Then click on **Next**.

7) Name your graph and your axes appropriately. You can also choose what other types of features you want to be included on your graph by accessing the tabs at the top of the **Chart Wizard** window. Click on **Next**.

8) Select the option of inserting your chart **As object in**: and select **Sheet 2**. Click on **Finish**.

9) Make appropriate adjustments to your chart size, font size, etc. to create a reasonable picture. Remember you can left click while your cursor is in any part of your chart to access formatting options for that particular region.

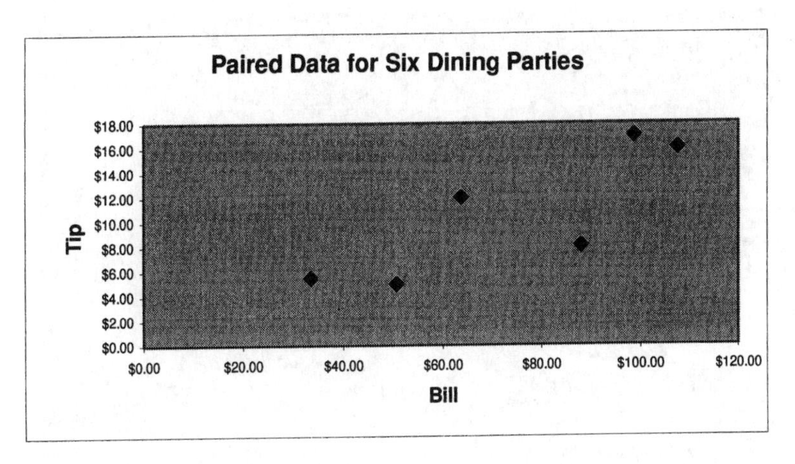

TO PRACTICE THESE SKILLS

You can apply the skills learned in this section by working on the following exercises.

1) Enter the data found in exercise 5 on page 520 of you textbook into Excel. Create the scatter plot, and find the linear correlation coefficient.

2) Load the data for Old Faithful Geyser from the CD that comes with your book. The file name is OLDFAITH.XLS. Copy the columns for Duration and Interval to a new worksheet. Create the scatter plot and find the linear correlation coefficient. Then copy the columns for Interval and Height to a new worksheet. Create the scatter plot and find the linear correlation coefficient. Use this information to address the question asked in exercise 11 on page 522 in your textbook.

3) Load the data on Diamond Prices, Carats and Color from the CD that comes with your book. The file name is DIAMOND.XLS. Use this data to address exercise 13 on page 522 of your textbook. You should copy the appropriate columns to a new worksheet for parts a and b.

SECTION 9-3 & 9-4: REGRESSION, VARIATION AND PREDICTION INTERVALS

In section 9-2 we concluded that there was evidence of linear correlation between the amount of the dinner bill and the tip that was left. Now we want to determine this relationship in order to be able to calculate the amount of the tip once we know the amount of the bill.

We have two options when working with linear regression, both of which are outlined below.

- The first option (**Add Trendline**) allows us to quickly generate the line of regression directly from our scatter plot.

- The second option uses the data analysis feature, and gives us a much more information, which will be useful in considering a more thorough analysis of the situation.

Option 1 – Add Trendline

1) Click anywhere in the **Chart** region, and then click on **Chart** on the main menu. Select **Add Trendline.**

2) Make sure that **Linear** is selected from the possible types shown.

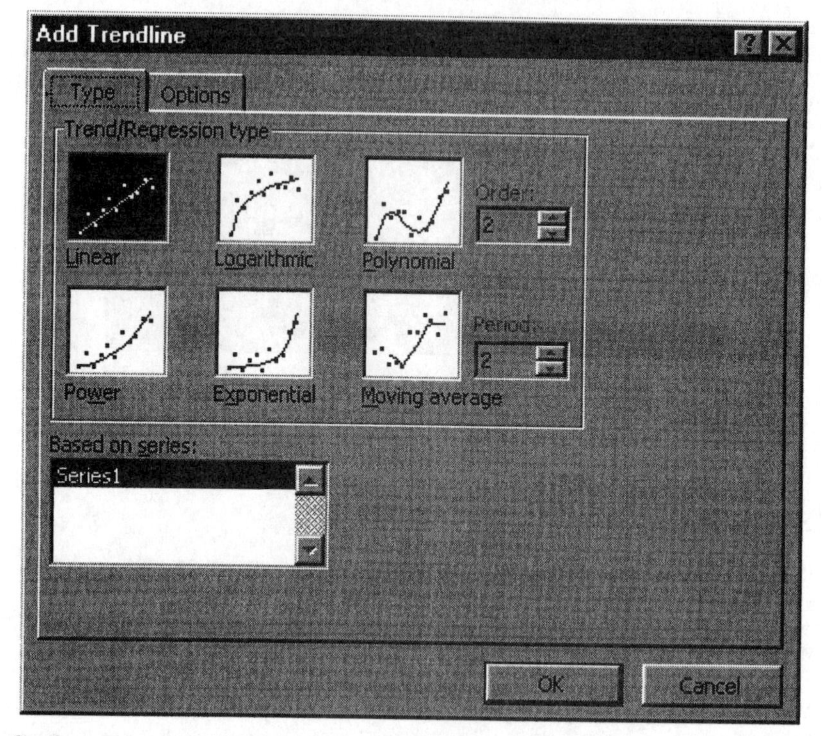

3) Click on the **Options** tab, and click in the box by **Display equation on chart**, and **Display R-squared value on chart.**

4) The trendline will automatically only include the beginning and ending output values from your data set. To extend the line, use the "Forecast" feature. In our example, we will want the trendline to go forward by 20 units, and backward by 30 units.

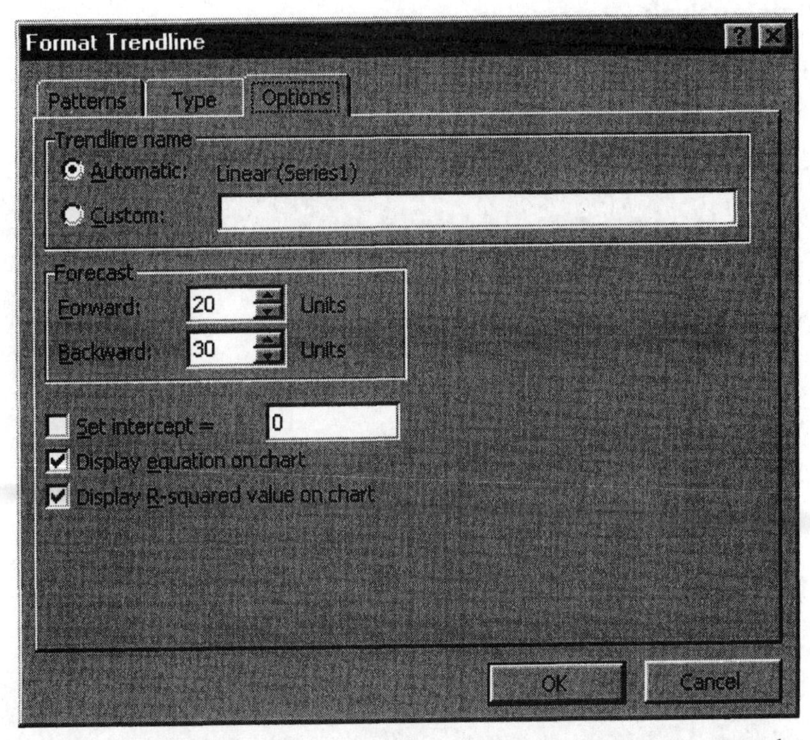

5) Click on **OK.** You will see your trendline and the equation superimposed on your scatter plot.

6) You will probably want to reformat your font size for the equation, and reposition where the equation appears on the screen so that it does not cover any of your data points. You can do this by selecting the region containing the equation, and when your cursor arrow shows **Series 1 Trendline 1 Equation,** double click to access the formatting menu. Again, with the arrow showing the **Series 1 Trendline 1 Equation** tag, hold down the left click button, and move the box where you want it to be within your plot area.

7) You may find that when you used the forecast feature, your original scaling on your scatter plot may have changed. You can make additional changes if desired by clicking on the area that you want to adjust, and making appropriate changes in the menu that appears.

8) Your picture should look similar to that shown below.

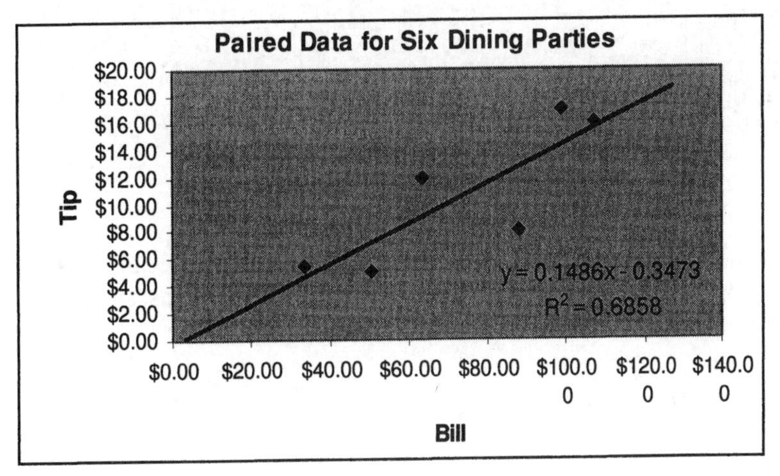

Option 2 – Data Analysis: Regression

1) Click on **Tools** from the main menu, then click on **Data Analysis**, and click on **Regression.** Click on **OK**.

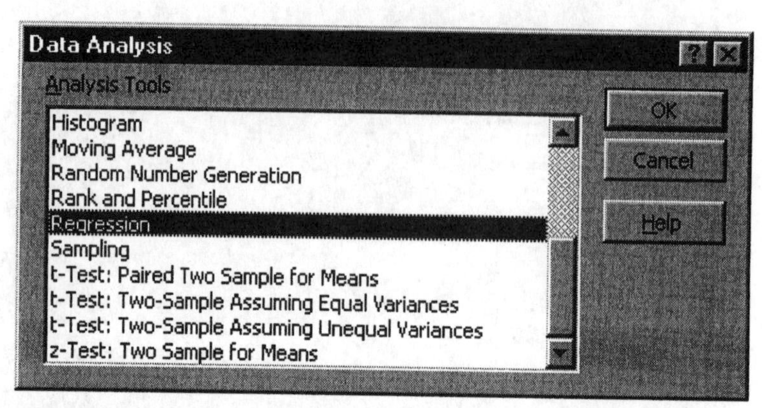

2) Complete the **Regression** dialog box as shown below.

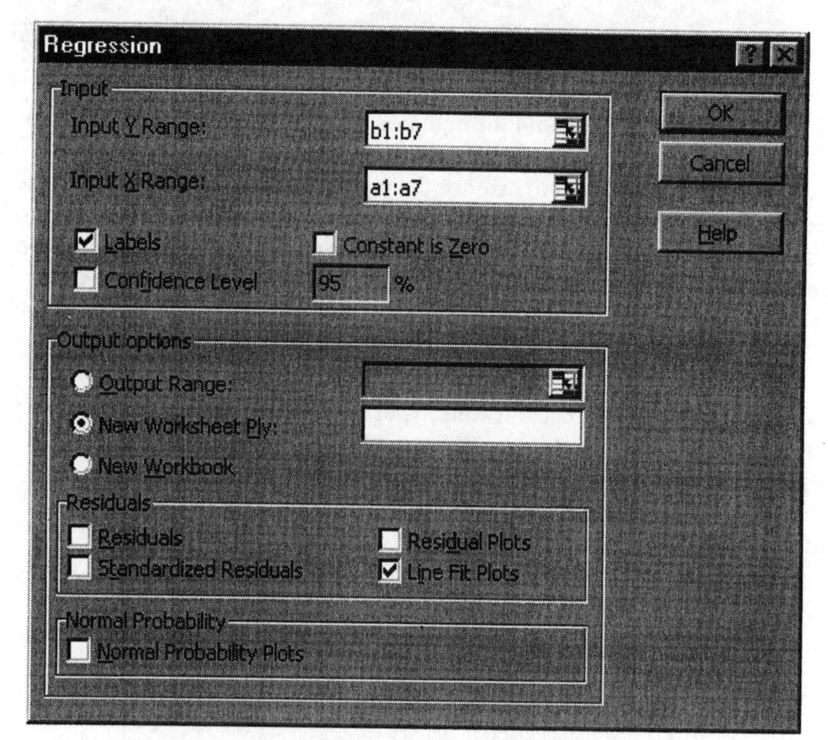

3) Click on **OK**.

4) You will be taken to a new sheet, and will need to resize the columns to see all the information provided clearly. Remember to do this, you can click on **Format,** click on **Columns,** and then click on **AutoFit Selection**. You will see the information shown on the next page.

SUMMARY OUTPUT

Regression Statistics	
Multiple R	0.8281591
R Square	0.6858476
Adjusted R Square	0.6073095
Standard Error	3.2658079
Observations	6

ANOVA

	df	*SS*	*MS*	*F*	*Significance F*
Regression	1	93.138329	93.138329	8.7326727	0.0417567
Residual	4	42.662004	10.665501		
Total	5	135.80033			

	Coefficients	*Standard Error*	*t Stat*	*P-value*	*Lower 95%*	*Upper 95%*
Intercept	-0.3472792	3.9360815	-0.0882297	0.9339348	-11.275616	10.581058
Bill	0.1486141	0.0502906	2.9551096	0.0417567	0.0089848	0.2882434

RESIDUAL OUTPUT

Observation	*Predicted Tip*	*Residuals*
1	4.6253502	0.8746498
2	7.1844858	-2.1844858
3	12.718877	-4.6388767
4	14.341743	2.6582568
5	9.1045806	2.8954194
6	15.604963	0.3950366

5) You will also see a scatter plot showing both the actual data points as well as the points generated from the regression equation. You will need to resize and reformat this graph to make it look the way you want it to. You will also need to add the actual regression line. To add the line, double click on any one of the **Predicted** points. In the **Format Data Series** box that comes up, click in the bubble in front of **Automatic** under **Line**. (This can be found under the **Patterns** tab option.) You should create a picture similar to that shown on the next page.

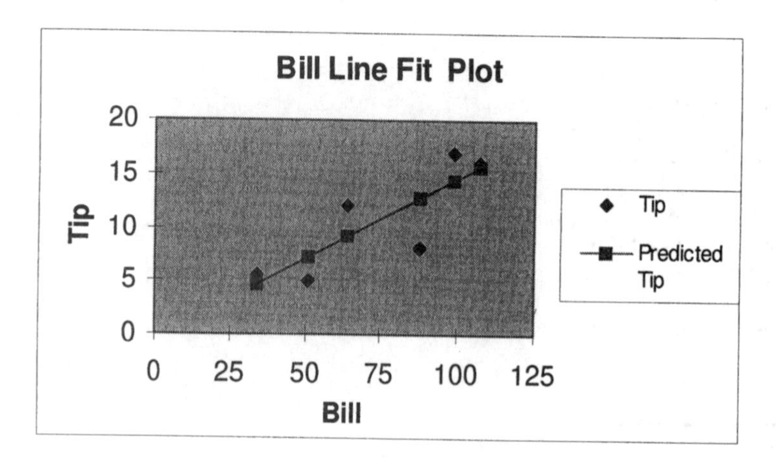

Interpreting This Output

Using the **Regression** option under **Data Analysis** provides you with more information than you need, but you can cut and paste the information that you need into another worksheet, or another document. To give you an idea of what the provided information represents, the major results are briefly described below:

- **Multiple R**: This is the correlation between the input variable (Bill) and the output variable (Tip). Since for this example there is only one input, the value given here is the correlation coefficient, r, expressing the linear relationship between the amount of the bill and the tip given.
- **R Square**: This is also referred to as the coefficient of determination. It represents the proportion of variation in the output that can be explained by its linear relationship with the input.
- **Adjusted R Square**: The sample R Square tends to be an optimistic estimate of the fit between the model and the real population. The adjusted R Square gives a better estimate.
- **Standard Error**: This is the standard error of the estimate, and can be interpreted as the average error in predicting the output by using the regression equation.
- **Observations**: This is the number of paired data values included in the analysis.
- **ANOVA**: You do not need to understand most of the information provided in this section for this chapter, but essentially this part of the output gives more detailed information about the variation in the output that is explained by the relationship with the input. For each source of variation, the output gives degrees of freedom (df), sum of squares (SS), the F value obtained by dividing the mean square (MS) regression by the mean square residual, and the significance of F, which is the P-value associated with the obtained value of F. A fuller treatment of the Analysis of Variance (ANOVA) can be found in chapter 11.
- **Coefficients**: These are the coefficients for your regression equation. The value listed in the first row is the y intercept of the regression line, while the value listed in the second row is the slope of the line.
- **T Stat**: This refers to a test of the hypotheses that the intercept is significantly different from zero.
- **P – Value:** This is the probability associated with the obtrained t statistic.
- **Lower and Upper 95%:** These are the confidence interval boundaries for both the intercept and the slope.
- **Residuals:** This table shows you the values that would be predicted for the output when using the regression equation for each input value. The second column shows the difference between the predicted value and the actual data value.

TO PRACTICE THESE SKILLS

You can practice the skills learned in this section by working on the following exercises.

1) Open the workbook where you saved the data from exercise 5 in section 9-2. Using this data, find the regression line utilizing both the **Trendline** option and the **Regression** option.

2) Open the workbook where you saved the data from exercise 11 in section 9-2 of your textbook. Using this data, find the regression lines and the other data using the **Regression** option for the paired data (duration, interval) and the paired data (height, interval). Use the information created under "Residual Output" to answer the questions asked in exercise 11 on page 536 of your textbook.

3) Open the workbook where you saved the data from exercise 13 in section 9-2 of your textbook. Using this data, find the regression lines and the other data using the **Regression** option for the paired data (price, carat) and the paired data (price, color). Use the information created under "Residual Output" to answer the questions asked in exercise 13 on page 536 of your textbook.

SECTION 9–5: MULTIPLE REGRESSION

The previous sections dealt with relationships between exactly two variables. This section presents a method for analyzing relationships that involve more than two variables. A multiple regression equation expresses a linear relationship between an output or dependent variable (y) and two or more inputs or independent variables (x values).

For this demonstration, we will use information on eight bears, as presented in Table 9-3 on page 550 of your textbook. We will create an equation which expresses weight as the output (y) and head length and total overall length as the input variables (x values).

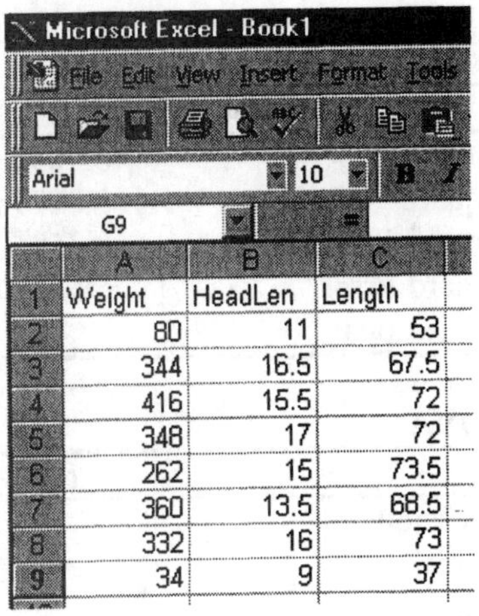

1) Enter the information given in Table 9-3 into a new Excel worksheet. You could also load the file BEARS.XLS from the CD that comes with your book, and copy the appropriate information on the 8 bears used into a new worksheet. Although the information shown in Table 9-3 is in rows, you can present your information in columns. **The values for the independent x values must be in adjacent columns.**

2) Click on **Tools, Data Analysis, Regression.**

Microsoft Excel - Book1

	A	B	C
1	Weight	HeadLen	Length
2	80	11	53
3	344	16.5	67.5
4	416	15.5	72
5	348	17	72
6	262	15	73.5
7	360	13.5	68.5
8	332	16	73
9	34	9	37

3) Fill in the **Regression** dialog box as shown, and click on **OK**.

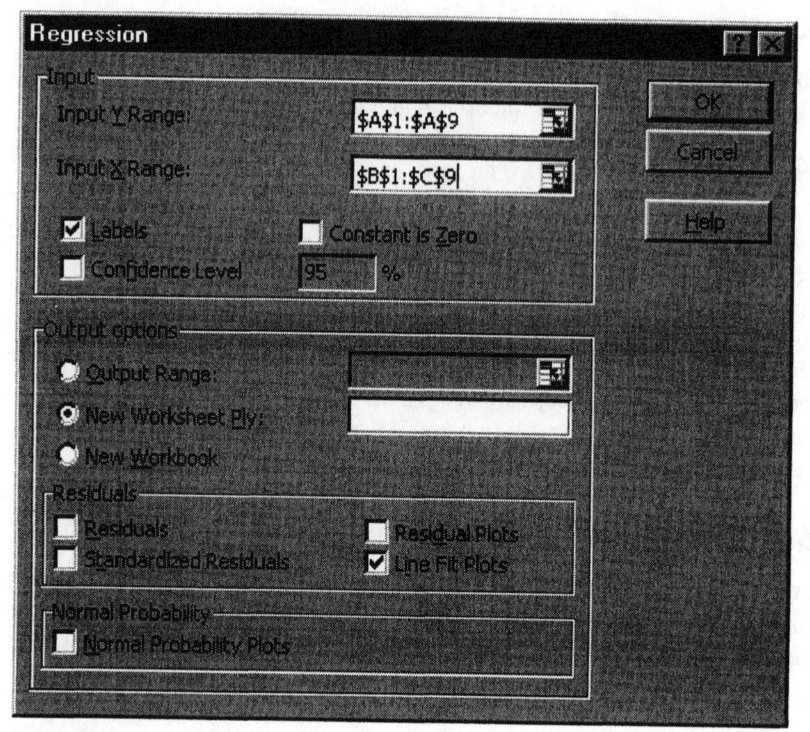

4) You will see the information given below.

SUMMARY
OUTPUT

Regression Statistics	
Multiple R	0.909935903
R Square	0.827983347
Adjusted R Square	0.759176686
Standard Error	68.56490584
Observations	8

ANOVA

	df	SS	MS	F	Significance F
Regression	2	113142.2684	56571.13422	12.03347661	0.012272308
Residual	5	23505.73157	4701.146313		
Total	7	136648			

	Coefficients	Standard Error	t Stat	P-value	Lower 95%
Intercept	-374.3034756	134.0930771	-2.791370619	0.03838268	-719.0001405
HeadLen	18.82040176	23.14805335	0.813044686	0.453152053	-40.68346649
Length	5.874757415	5.065488967	1.159761171	0.298513005	-7.146475235

5) From this information, you can create the equation: Weight = - 374.3 + 18.8(HeadLen) + 5.87(Length). For a discussion on other important elements, refer to the discussion in section 9-5 of your book.

TO PRACTICE THESE SKILLS

You can practice the skills learned in this section by working on the following exercises.

1) Use the data from the file BEARS.XLS in its entirety to answer the questions posed for exercises 5 and 7 on page 557 of your textbook.

2) Use the data from the file DIAMONDS.XLS. Since you have worked with this data in previous sections, you may already have a file where this data is loaded into one of the sheets. If so, you can add sheets to this file to complete exercises 13 and 15 on page 557 of your textbook.

CHAPTER 10: MULTINOMIAL EXPERIMENTS AND CONTINGENCY TABLES

SECTION 10-1: OVERVIEW

In earlier chapters you learned that the first step in organizing and summarizing data for a single variable was to create a frequency table. It is often desirable to categorize data according to two quantitative variables for the purpose of determining whether or not these variables are related. This data is organized by using a **cross classification table** or a **contingency table.** You have already learned the basics for creating tables in Excel by using the **Pivot Table** wizard introduced in Chapter 3. We will look at the Chi Square test for Independence, used to determine whether a contingency table's row variable is independent of its column variable.

CHITEST: returns the test for independence: the value from the chi-squared distribution for the statistic and the appropriate degrees of freedom.

CHITEST(actual_range, expected_range) where the actual_range is the range of data that contains observations to test against expected values and the expected_range is the range of data that contains the ratio of the product of row totals and column totals to the grand total.

SECTION 10 - 2: MULTINOMIAL EXPERIMENTS:GOODNESS-OF-FIT

In previous chapters you looked at several different hypothesis tests that assumed that the data came from a normally distributed population. There are other less formal ways to check to see if a population is normally distributed. These might include creating a histogram and observing if the shape resembles a normal distribution. While not a bad approach, we will feel more secure about decisions we make if we can substantiate our findings with a formal statistical technique. A **goodness-of-fit test** is one such technique.

While much of the work presented here can be done using the traditional paper and pencil approach, the use of technology makes the computational aspect of this problem much easier. By entering a formula in a cell and then copying that formula throughout the appropriate cells, it is possible to save time and avoid arithmetic mistakes.

TO PRACTICE THESE SKILLS

You can practice the skills covered in this section by working through the following problems.

1) Set up an Excel spreadsheet with the same headings found on page 580 of your textbook. Use Excel to work through problem 9 on page 586. This problem is very similar to the example found in Table 10-3 on page 580.

2) Use the data found in Data Set 17 in Appendix B of your textbook or on the CD data disk in the file "BOSTRAIN" to work through problem 15 on page 587 of your textbook. You will want to set up an Excel spreadsheet similar to the one set up in the used above.

SECTION 10-3: CONTIGENCY TABLES: INDEPENDENCE AND HOMOGENEITY

Excel provides the tools necessary to do a chi-square test for independence, although they are not found within a single analysis tool. The process involves creating a contingency table using the **Pivot table** command in Excel, creating a table of expected and observed frequencies and using the **CHITEST** function.

The contingency table is the backbone of the chi-square test for independence in Excel. Assume that we had entered the necessary information for the survival figures of the passengers of the Titanic and were able to create the pivot table below. This is a duplicate of Table 10-1 found in the chapter problem at the beginning of chapter 10.

	A	B	C	D	E	F	H	I
1			Gender/Age Category					
2			Men	Women	Boys	Girls	Row Totals	
3	Survived		332	318	29	27	706	
4	Died		1360	104	35	18	1517	
5								
6	Column Totals		1692	422	64	45		

CREATING A TABLE OF EXPECTED FREQUENCIES

The contingency table provides the actual frequencies for each cell. To perform the chi-squared test for independence we also need the expected frequencies. Excel will expect to find these expected frequencies in a separate table and not within the pivot table created.

1) Begin by copying the row and column headings from the pivot table to a location just below it in your Excel worksheet. We will use this new table for the expected frequencies.

2) Using the formula **expected frequencies** = $\frac{(row\ total)(column\ total)}{grand\ total}$ and the appropriate cell addresses we complete the table created in step (1).

Note:

You must use relative and absolute references if you are going to copy your formula to all of your cells and not just retype them in. If you have forgotten how to use relative and absolute references see Chapter 1, Section 1-6, for help.

---

3) If all is done properly you should see

	Expected Frequencies			
	Men	Women	Boys	Girls
Survived	537.360	134.022	20.326	14.291
Died	1154.640	287.978	43.674	30.709

PERFORMING THE CHI-SQUARE TEST

The function statistician function **CHITEST** performs the chi-square test. This function asks for the observed and expected values and will return the p value of the test.

1) Click on **Insert**, highlight **Function** and click.

2) From the **Paste Function Dialog box** highlight **Statistical** and then highlight **CHITEST.**

3) Click **OK.**

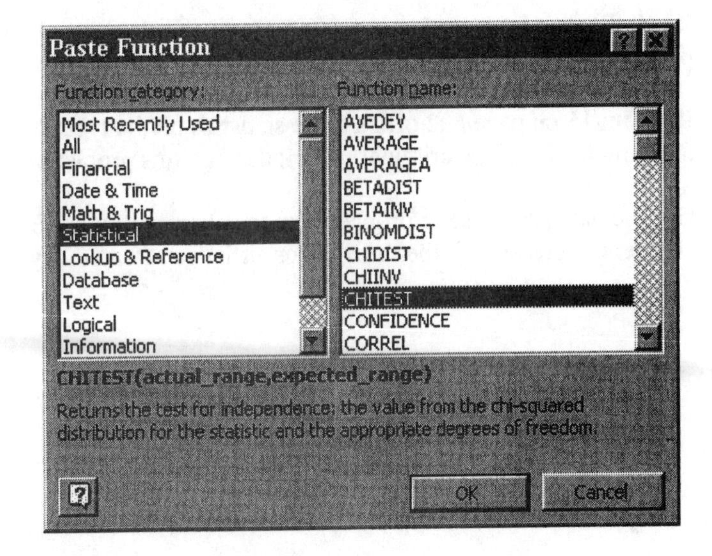

4) The **CHITEST** dialog box opens as shown below.
   a) For the **Actual_range** highlight those cells that contain the observed frequencies. Be careful not to highlight the row and column totals.
   b) For the **Expected_range** highlight those cells that contain the expected frequencies.

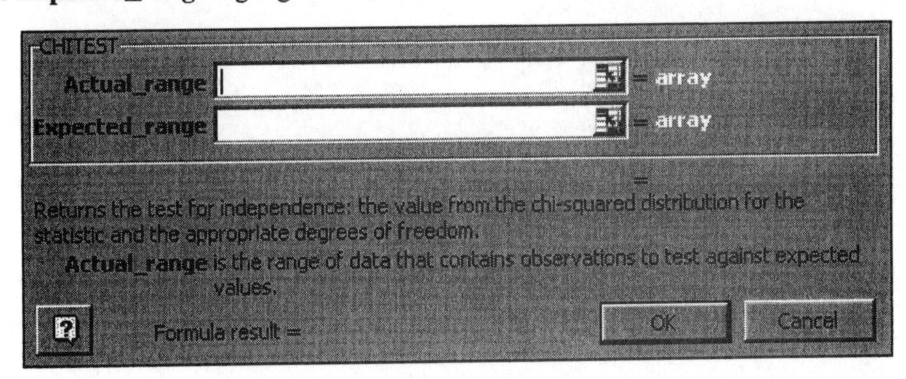

5) Click **OK.**

This returns a very small P-value. Therefore we reject the null hypothesis stated in the original problem found on page 593. That is, we reject the null hypothesis that whether a person survived the sinking of the Titanic is independent of whether the person is a man, woman, boy or girl.

TO PRACTICE THESE SKILLS

You can practice the skills covered in this section by working through the following problems found in your textbook.

1) Work through problem 7 on page 600 to test the claim that there is a gender gap in the confidence that people have in the police.

2) Problem 15 on page 602 offers some statistics on the type of crime committed and its relationship to the drinking habits of the criminal. Work through this problem using the method outlined in this section.

3) Use the data presented in problem 4 in the chapter Review Exercises and the methods presented in this chapter to determine if the three major airlines presented have the same proportion of on-time flights.

CHAPTER 11: ANALYSIS OF VARIANCE

SECTION 11-1: OVERVIEW

In this chapter, we consider a procedure for testing the hypothesis that three or more means are equal. We will use the Analysis of variance (ANOVA) features of Excel.

Excel tools introduced in this section are outlined below.

ANOVA SINGLE FACTOR

This feature returns summary statistics on the data, as well as Analysis of Variance information.

ANAOVA: TWO FACTOR WITH REPLICATION

This feature returns summary statistics for each group in your data set, as well as Analysis of Variance information.

SECTION 11–2: ONE-WAY ANOVA

1) Enter the data for Head Injuries from table 11-1 on page 614 of your textbook into Excel as shown below.

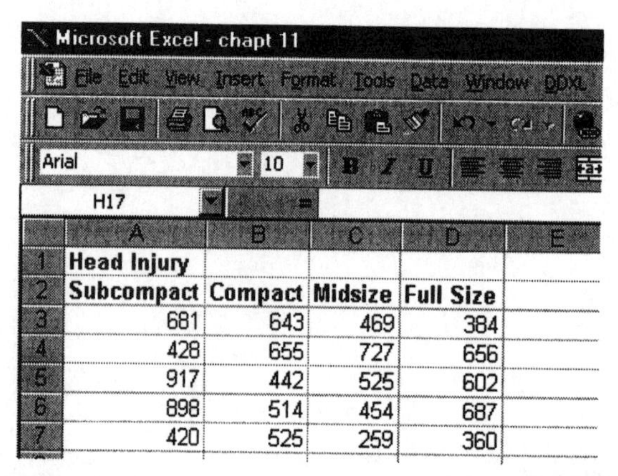

	A	B	C	D	E
1	Head Injury				
2	Subcompact	Compact	Midsize	Full Size	
3	681	643	469	384	
4	428	655	727	656	
5	917	442	525	602	
6	898	514	454	687	
7	420	525	259	360	

2) Click on **Tools** and click on **Data Analysis.** Click on **Anova: Single Factor**, and click on **OK.**

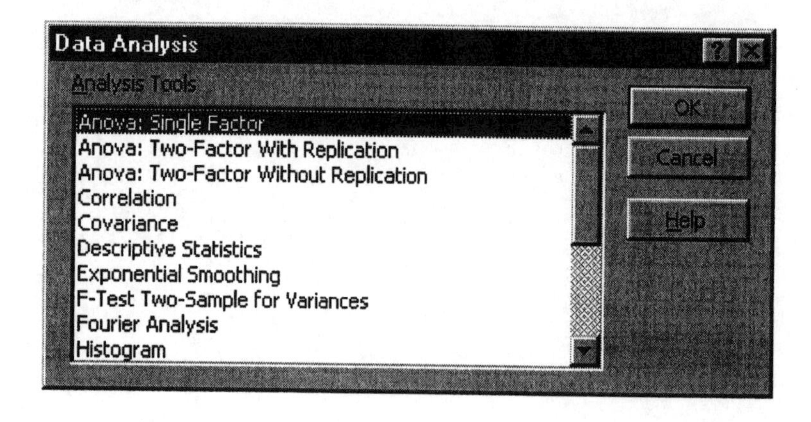

3) In the dialog box, type in "$A$2:$D$7" in the **Input Range**. Make sure that **Columns** and Labels in First Row are selected, and that **Alpha** is set at 0.05. You should have your results appear in a new worksheet. Then click on **OK**.

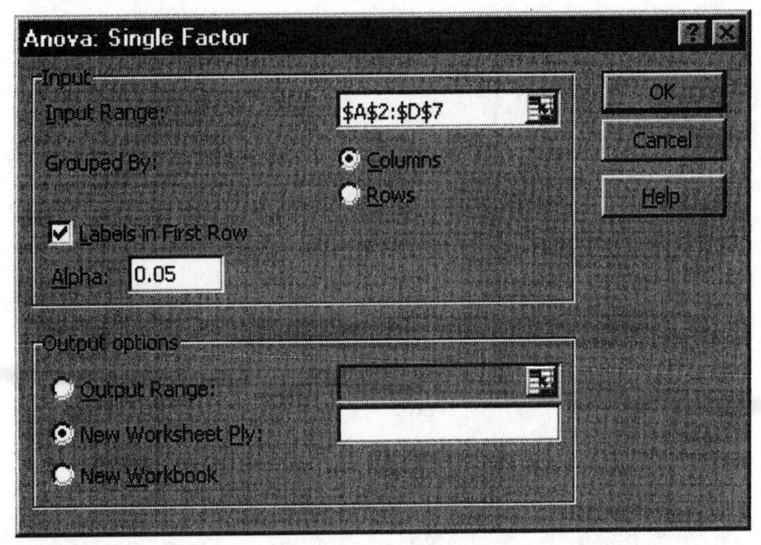

4) You will see a table of values that is automatically selected. Click on **Format, Column**, and click on **AutoFit Selection**.

5) You will see the table below.

Anova: Single Factor

SUMMARY

Groups	*Count*	*Sum*	*Average*	*Variance*
Subcompact	5	3344	668.8	58542.7
Compact	5	2779	555.8	8272.7
Midsize	5	2434	486.8	28110.2
Full Size	5	2689	537.8	23905.2

ANOVA

Source of Variation	*SS*	*df*	*MS*	*F*	*P-value*	*F crit*
Between Groups	88425	3	29475	0.992167014	0.421569916	3.238866952
Within Groups	475323.2	16	29707.7			
Total	563748.2	19				

6) For a discussion of the key components of this information, see the material on pages 622 – 624 of your book.

TO PRACTICE THESE SKILLS

You can practice the skills learned in this section by working on the following exercises.

1) Follow through the directions from this section to complete exercise 5 on page 627 of your textbook.

2) Load the data from the M&M.XLS file on the CD that comes with your book, or open the file where you have previously worked with this data. Use the techniques presented in this section to complete exercise 11 on page 628 of your textbook.

SECTION 11-3: TWO WAY ANOVA

For this demonstration, we will use the information presented in table 11–4 on page 631 of your textbook. You **MUST** enter the information for females and males down a column, not across a row.

1) Enter the data in Excel as shown below.

	Female	Male
Red	1130	1257
Red	621	898
Red	813	743
Red	996	921
Red	1030	1179
Red	1092	1133
Red	855	896
Red	896	1190
Red	858	908
Red	1095	699
Green	996	993
Green	630	1025
Green	583	907
Green	828	1111
Green	1121	1147
Green	780	1180
Green	916	1229
Green	793	1450
Green	1188	1071
Green	499	1153
Blue	706	1611
Blue	1068	939
Blue	1013	1004
Blue	892	821
Blue	1370	915
Blue	866	1244
Blue	848	996
Blue	1408	1131
Blue	793	1039
Blue	1097	1159

2) Click on **Tools, Data Analysis,** and then click on **ANOVA: Two-Factor with Replication**. Click on **OK**.

3) Type in "$A$1:$C$31" in the box for **Input Range**. Type in 10 for **Rows per sample** and 0.05 for **Alpha**. You should have your results appear in a new worksheet. Click on **OK**.

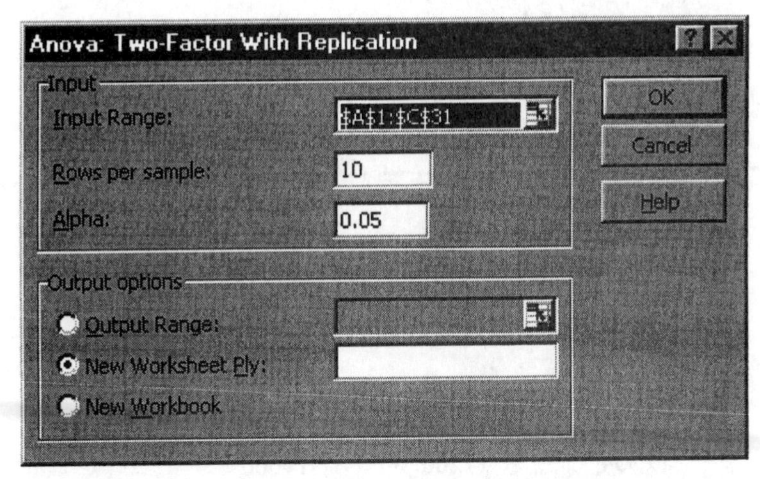

4) You will need to format the columns so that the titles all show up fully. You will see the data that is presented on the next page. For a detailed discussion of the values in this table, see section 11-3 in your textbook.

TO PRACTICE THESE SKILLS

You can practice the skills learned in this section by completing the following exercise.

Load the data file STATSURV.XLS from the CD that comes with your book. From this data, select nine values to work with according to exercise 11 on page 640 of your textbook.

Anova: Two-Factor With Replication

SUMMARY	Female	Male	Total
Red			
Count	10	10	20
Sum	9386	9824	19210
Average	938.6	982.4	960.5
Variance	25357.82222	37928.48889	30482.57895
Green			
Count	10	10	20
Sum	8334	11266	19600
Average	833.4	1126.6	980
Variance	51347.15556	22085.37778	57406.52632
Blue			
Count	10	10	20
Sum	10061	10859	20920
Average	1006.1	1085.9	1046
Variance	55466.98889	49561.21111	51426
Total			
Count	30	30	
Sum	27781	31949	
Average	926.0333333	1064.966667	
Variance	46242.86092	37817.82644	

ANOVA

Source of Variation	SS	df	MS	F	P-value	F crit
Sample	80310	2	40155	0.99662	0.37581	3.16824
Columns	289537.0667	1	289537.0667	7.18611	0.00972	4.01954
Interaction	181726.5333	2	90863.26667	2.25516	0.11464	3.16824
Within	2175723.4	54	40291.17407			
Total	2727297	59				

CHAPTER 12: STATISTICAL PROCESS CONTROL

SECTION 12 – 1 OVERVIEW

In this chapter, we address changing characteristics of data over time. In monitoring this characteristic, we are able to control the production of goods and services.

The major features used in this section are ones that have already been introduced in earlier sections, and include:

Chart Wizard to create a line graph from a set of data points.

Function used to access **Average, Median,** and **STDEV.**

The new feature introduced in this section is how to use the **Callouts** under the Draw menu.

SECTION 12–2: CONTROL CHARTS FOR VARIATION AND MEAN

In this section, we will consider data arranged according to some time sequence. We will consider the information on Axial Loads (in pounds) of Aluminum Cans found in Table 12-1 on page 653 of your text.

1) Load the file CANS.XLS from the CD that comes with your textbook, and copy the information on Aluminum cans 0.0109 in. Load (pounds) to a new worksheet. Your worksheet should look like the first part of Table 12-1 in your textbook.

2) To create the columns for the mean, median and sample standard deviation for each row, click on the function icon and select **Statistical, Average (or Median, or STDEV)**. Click on **OK**, and in the dialog box, enter cells B2 through H2. Press **Enter**. To create two decimal places for the Mean and Standard Deviation, click on **Format, Cells, Number**, and type in 2 for the number of decimal places you want to have shown. Use the fill handle to fill in the rest of the column.

3) To create the column for the Range, position your cursor in cell K2, and type in the formula: = Max(b2:h2)-Min(b2:h2). Press **Enter**. Use the fill handle to fill in the rest of the column.

4) You should now have the table shown on the next page.

Day								Mean	Median	Range	s
1	270	273	258	204	254	228	282	252.71	258	78	27.63
2	278	201	264	265	223	274	230	247.86	264	77	29.66
3	250	275	281	271	263	277	275	270.29	275	31	10.56
4	278	260	262	273	274	286	236	267.00	273	50	16.34
5	290	286	278	283	262	277	295	281.57	283	33	10.72
6	274	272	265	275	263	251	289	269.86	272	38	11.84
7	242	284	241	276	200	278	283	257.71	276	84	31.45
8	269	282	267	282	272	277	261	272.86	272	21	7.90
9	257	278	295	270	268	286	262	273.71	270	38	13.45
10	272	268	283	256	206	277	252	259.14	268	77	25.87
11	265	263	281	268	280	289	283	275.57	280	26	10.10
12	263	273	209	259	287	269	277	262.43	269	78	25.29
13	234	282	276	272	257	267	204	256.00	267	78	27.81
14	270	285	273	269	284	276	286	277.57	276	17	7.32
15	273	289	263	270	279	206	270	264.29	270	83	26.98
16	270	268	218	251	252	284	278	260.14	268	66	22.26
17	277	208	271	208	280	269	270	254.71	270	72	32.15
18	294	292	289	290	215	284	283	278.14	289	79	28.13
19	279	275	223	220	281	268	272	259.71	272	61	26.47
20	268	279	217	259	291	291	281	269.43	279	74	25.87
21	230	276	225	282	276	289	288	266.57	276	64	27.21
22	268	242	283	277	285	293	248	270.86	277	51	19.32
23	278	285	292	282	287	277	266	281.00	282	26	8.41
24	268	273	270	256	297	280	256	271.43	270	41	14.26
25	262	268	262	293	290	274	292	277.29	274	31	14.08

Creating a Control Chart for Monitoring Means

1) Before we create the actual chart for the means, we need to compute the value for the Centerline, and the values for the Upper Control Limit (UCL) and the Lower Control Limit (LCL).

2) To find the mean of the sample means, position your cursor in a cell under the column containing the sample means. Click on the **Function** icon, click on **Statistical** and **Average.** Select the cells containing the sample means, and then press **Enter**. You should find the mean of the sample means is 267.11.

3) To find the upper and lower control limits, we must use table 12–2 on page 660 and locate the value for A_2. Since the number of observations in each subgroup is 7, we look under the column for the mean, and find that A_2 is 0.419.

4) We must also find the mean of the ranges. You can copy the formula from the cell containing the mean of your sample means to a cell under the column showing the ranges. You should find that the mean of the range values is 54.96.

5) Compute the upper control limit using the following pattern: Mean of sample means + A_2 * (mean of the ranges). Your upper control limit will be 267.11+ (.419)(54.96) = 290.14

6) Compute the lower control limit using the following pattern: Mean of sample means + A_2 * (mean of the ranges). Your lower control limit will be 267.11 – (.419)(54.96) = 244.09.

7) We need to add these values to our table. Select the three columns **directly to the right** of the column containing your sample means by positioning your cursor on the column letter, holding the mouse down, and dragging over the other two columns so that all three are highlighted. Click on **Insert** and select **Columns.** You should now have three blank columns inserted in your table directly to the right of the column of means.

8) At the top of the first blank column, type in “Mean of means”. In the cell directly underneath this, type in the value you produced above of 267.11. Use the fill handle to fill in the remaining rows of this column with this value.

9) At the top of the next column, type LCL. In the cell directly underneath this, type in the value of your lower control limit, and fill the remaining table rows in this column. Repeat this procedure for the upper control limit.

10) Click on the **Chart Wizard** icon. For the **Chart type**, click on **Line**, and for the **Chart subtype**, click on the first graph in the second row. Then click on **Next**.

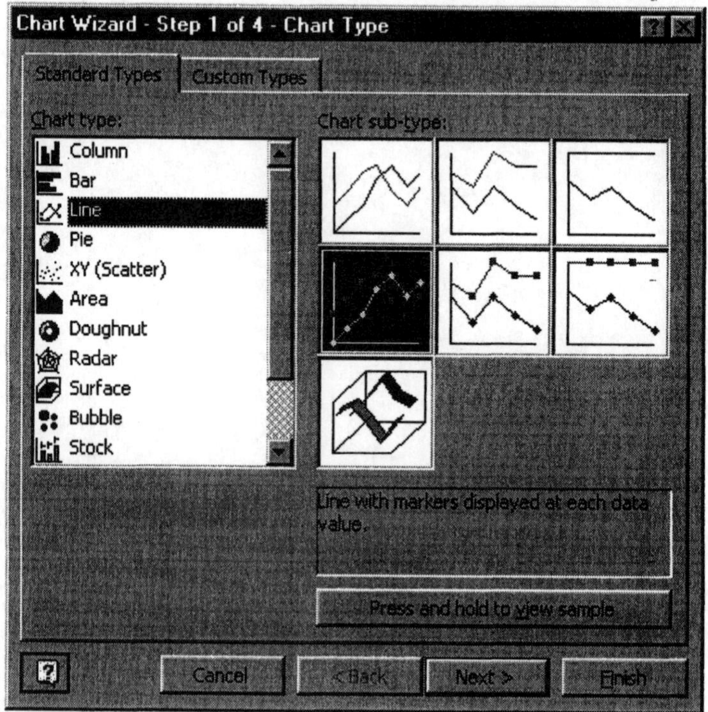

11) Make sure that the bubble by **Columns** is checked. Select cells I1 through L26. You will see =Sheet1!$I$1:$L$26 entered in the box by **Data Range**. Your dialog box should now look like the one shown below. Click on **Next.**

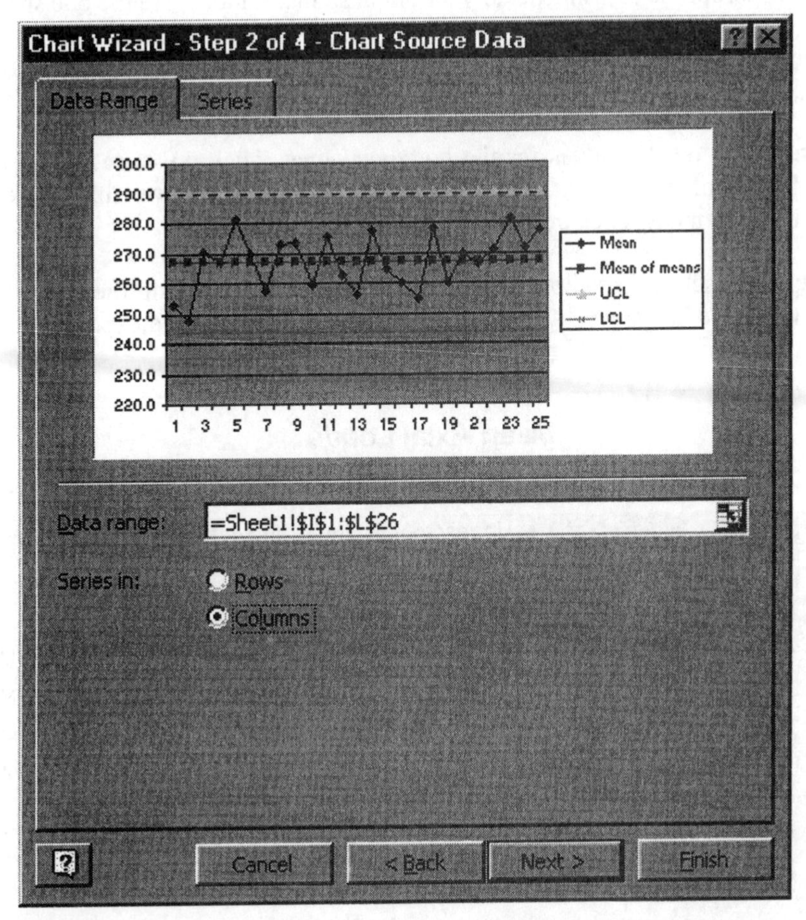

12) Fill in the **Chart Options** as shown. Then click on the **Gridlines** tab, and make sure that none of the options are selected. Then click on **Next**.

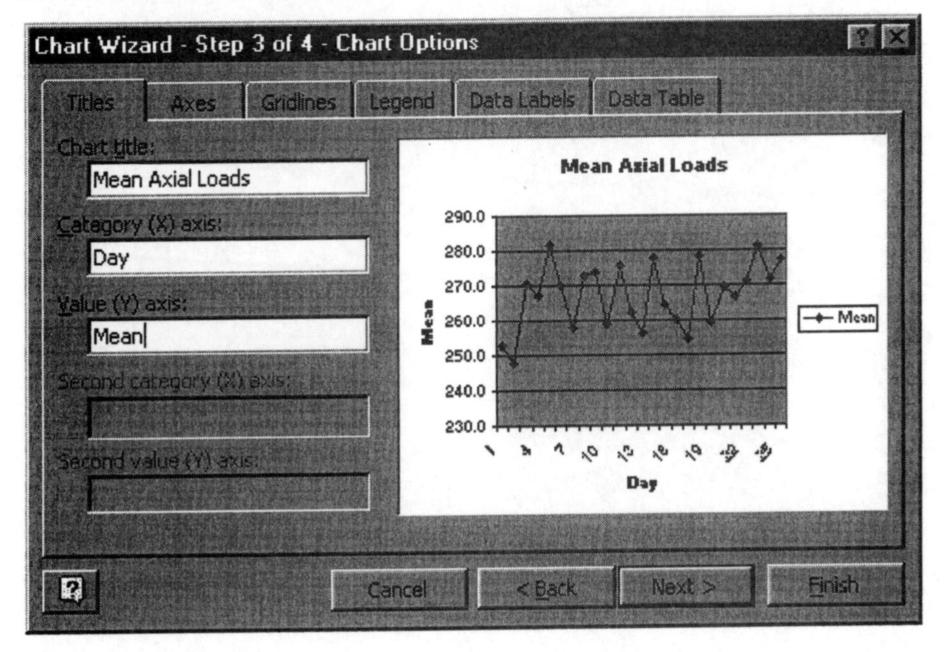

13) For **Chart Location**, select **As object in**, and choose **Sheet 2** from the drop down menu. Then click **Finish**.

14) You should make appropriate adjustments to your graph so that it looks like the one shown below.

15) To eliminate the markers on the horizontal lines, click on any one of the markers. In the **Format Data Series** dialog box, click in the bubble beside **None** in the column labeled **Markers.**

16) To add the information on the values for the horizontal lines, click on **AutoShapes** on the **Drawing Toolbar** at the bottom of your screen. (If the Drawing Toolbar is not available, click on **View,** select **Toolbars**, and from this menu, click in front of **Drawing**.)

17) Click on **Callouts**, and select the callout style of your choice. You can edit the font size, and move the box where you want it to be, as well as shifting the "handle". You should experiment with this option, as it provides an excellent way to add information to any graph.

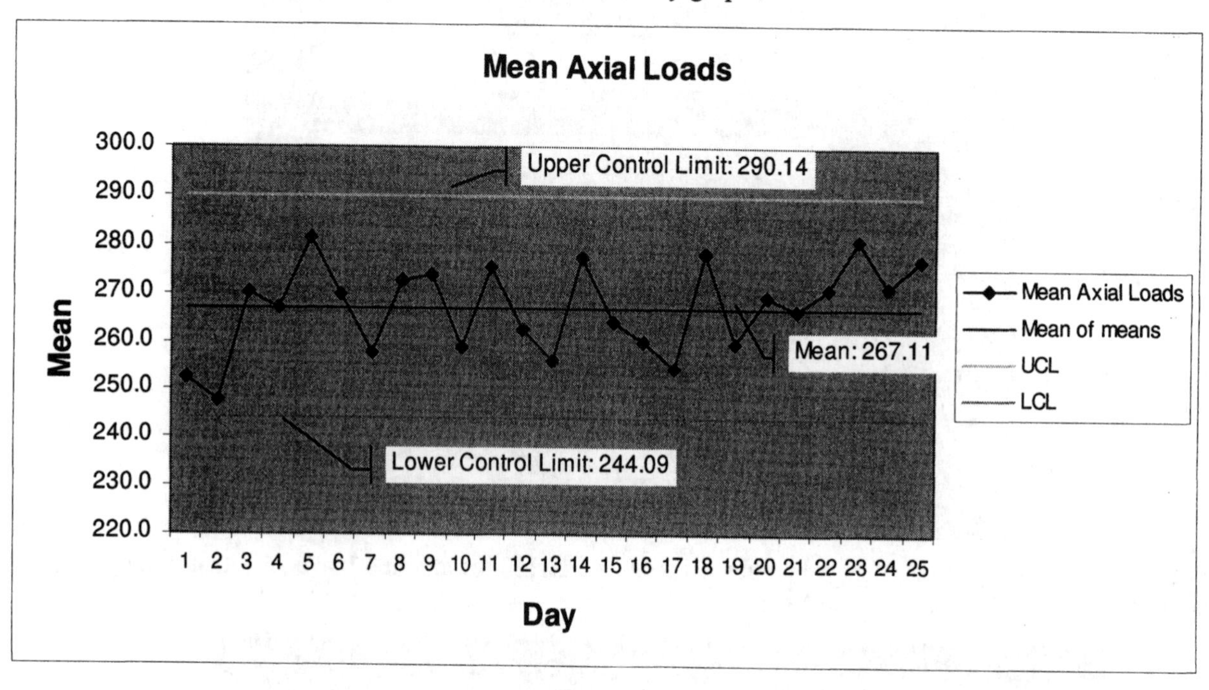

Control Chart for Monitoring Variation: The R Chart

You can follow basically the same procedures as those above to create an R chart (or range chart). The major differences are outlined below.

1) Your **Data Range** and **Chart Title** will need to reflect the fact that you are using the sample ranges, not the sample means.

2) Instead of using the formulas presented in the previous directions, you will compute your Lower Control Limit by using the formula: $D_3(\overline{R})$ and the Upper Control Limit by using the formula $D_4(\overline{R})$. The values of D can be found in Table 12-2 on page 660 of your text.

3) $\overline{R}$ can be computed using Excel, and represents the mean of the sample ranges.

4) You will need to create columns showing the values for $\overline{R}$ and the Upper and Lower Control Limits next to your column for Range in order to create the appropriate horizontal lines on your chart.

5) You will need to rearrange your data so that the columns containing the ranges, the $\overline{R}$ value, and the Upper and Lower Control Limits are adjacent to each other. You can do this by either inserting additional columns next to the column containing the ranges, or by copying your range column to another part of the worksheet, leaving you room to add the additional columns necessary.

6) Your final graph should end up looking much like the one shown below. Again, you could add the values for your $\overline{R}$, UCL and LCL by utilizing the **Callout** feature in the **AutoShapes** menu.

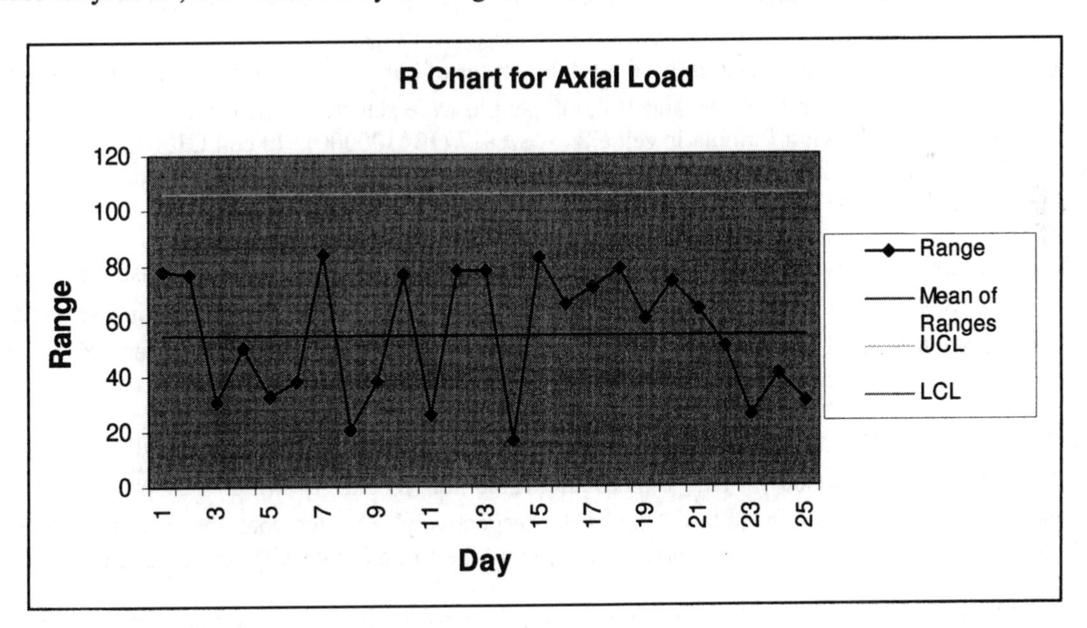

TO PRACTICE THESE SKILLS

You can apply the skills learned in this section by working on the following exercises.

1) Load the data from the file BOSTRAIN.XLS from the CD that comes with your book. Use this data to complete exercise 11 on page 668 of your textbook.

2) Using the data from BOSTRAIN.XLS, complete exercise 12 on page 668 of your textbook.

SECTION 12-3: CONTROL CHARTS FOR ATTRIBUTES

Whereas in section 12-2 we worked with quantitative data, this section works with qualitative data. We will again be selecting samples of size n at regular time intervals and plot points in a sequential graph with a centerline and control limits.

For illustration, we will work with the information on the example **Deaths from Infectious Diseases** on page 669 of your text.

1) Enter the information on number of deaths in column A of a new worksheet.

2) You need to create a column showing the proportion of the sample that each number of deaths represents. Type "p" in cell B1. Position your cursor in cell B2. Since each year, 100000 people were selected, we will enter the formula: =A2/100000 and press **Enter.** You should see the value .00025 in cell B2. Copy this formula down through cell B14.

3) Since we need to find $\bar{p}$, we need to find the sum of the values in column A. Position your cursor in cell A16, and type in the word "sum". Move to cell A17, and select the **Summation** symbol on your toolbar. You will see the formula: = sum(A2:A16), and you will see that the cells A2 through A16 have been selected. Click to the right of the 6 within the parentheses, delete this number, and type in 4, since you only want to add from A2 through A14. Then click **Enter**. You should now see the value 375 listed.

4) To find $\bar{p}$, you need to divide this sum by the total number of subjects sampled. Since the example is for information obtained over 13 years, and 100,000 people were selected each year, you can create this value by creating the following formula in cell C2: = $A$17/(13*100000). In cell C1, name the column "p-bar". Then, since you are going to want to graph a horizontal line at this height on your graph, copy the value obtained from the formula (.000288) down through cell C14.

5) You now need to compute the upper control limit and the lower control limit. We will use the formulas found on page 669 of your text. To find $\bar{q}$, position your cursor in cell C16 and type in the word "q-bar". Then move to cell C17 and enter the following formula: =1-$C$2. Then press **Enter**. You should see the value 0.999712 listed.

6) Type "LCL" in cell D1 of your worksheet, and enter the following formula in cell D2: =$C$2-3*SQRT($C$2*$C$17/(100000)) This utilizes the value for $\bar{p}$ and $\bar{q}$ which has previously been computed in your worksheet in cells C2 and C17 respectively. Notice that the cell addresses are absolute, meaning that they will not be updated as we copy the formula into different columns.

7) Type "UCL" in cell E1, and copy the formula from cell D2 into cell E2. Position your cursor in front of the minus sign in the formula bar, delete it, and type in +. Then press **Enter**. You should see the value .00045 in cell E2. Copy this formula down through cell E14.

8) When you are done with this work, you should have a table much like that shown on the next page.

# of Deaths	p	p-bar	LCL	UCL
25	0.00025	0.000288	0.000127	0.00045
24	0.00024	0.000288	0.000127	0.00045
22	0.00022	0.000288	0.000127	0.00045
25	0.00025	0.000288	0.000127	0.00045
27	0.00027	0.000288	0.000127	0.00045
30	0.0003	0.000288	0.000127	0.00045
31	0.00031	0.000288	0.000127	0.00045
30	0.0003	0.000288	0.000127	0.00045
33	0.00033	0.000288	0.000127	0.00045
32	0.00032	0.000288	0.000127	0.00045
33	0.00033	0.000288	0.000127	0.00045
32	0.00032	0.000288	0.000127	0.00045
31	0.00031	0.000288	0.000127	0.00045
Sum		q-bar		
375		0.999712		

9) Now use your **Chart Wizard**, and **Line** graph option to create the Control Chart. You will want to use a **Data Range** of cells B1 through E 14. Make sure that you deselect any gridline options. After modifying your graph as appropriate, you should end up with a picture similar to that shown below.

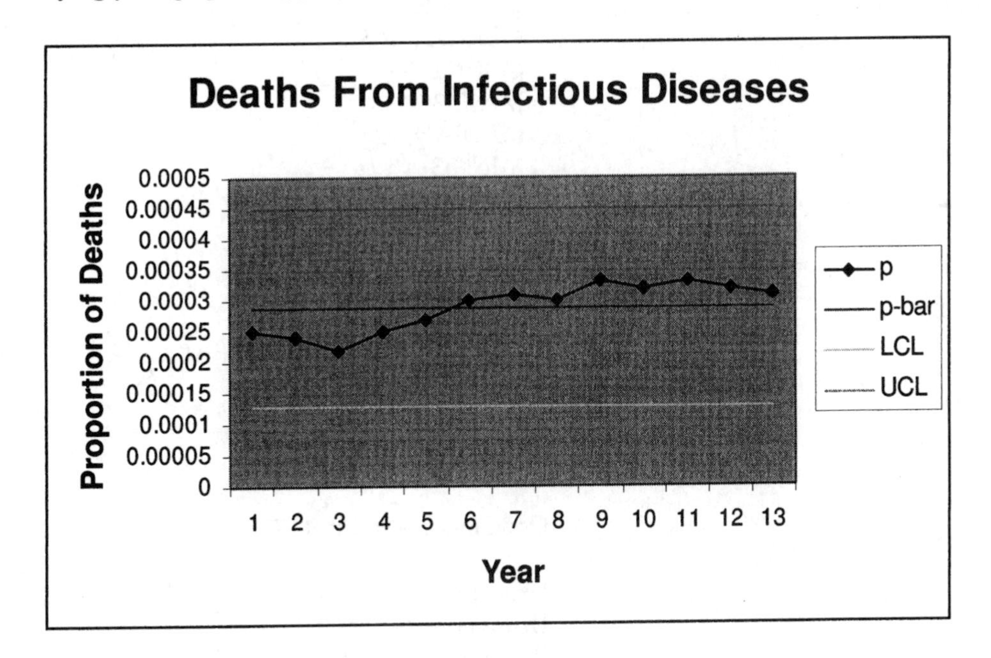

TO PRACTICE THESE SKILLS

You can apply the technology skills learned in this section by completing the following exercise.

Load the data from the file BOSTRAIN.XLS on the CD that comes with your book, or open a file where you have already retrieved this data. Delete the 53rd value for Wednesday. Compute the sample proportion for each of the 52 weeks by setting up a ratio of the number of days it rained divided by the 7 days of the week. Then complete exercise 7 on page 673 of your textbook.

INDEX